Diabetes Mellitus in the Elderly

Diabetes Mellitus in the Elderly has been co-published simultaneously as *Journal of Geriatric Drug Therapy*, Volume 12, Number 2 1999.

The *Journal of Geriatric Drug Therapy* Monographic "Separates"

Below is a list of "separates," which in serials librarianship means a special issue simultaneously published as a special journal issue or double-issue *and* as a "separate" hardbound monograph. (This is a format which we also call a "DocuSerial.")

"Separates" are published because specialized libraries or professionals may wish to purchase a specific thematic issue by itself in a format which can be separately cataloged and shelved, as opposed to purchasing the journal on an on-going basis. Faculty members may also more easily consider a "separate" for classroom adoption.

"Separates" are carefully classified separately with the major book jobbers so that the journal tie-in can be noted on new book order slips to avoid duplicate purchasing.

You may wish to visit Haworth's website at . . .

http://www.haworthpressinc.com

. . . to search our online catalog for complete tables of contents of these separates and related publications.

You may also call 1-800-HAWORTH (outside US/Canada: 607-722-5857), or Fax 1-800-895-0582 (outside US/Canada: 607-771-0012), or e-mail at:

getinfo@haworthpressinc.com

Diabetes Mellitus in the Elderly, edited by James W. Cooper, PharmPhD (Vol. 12, No. 2, 1999). *Discusses the diagnosis and treatment of diabetes mellitus in the elderly.*

Gastrointestinal Drug Therapy in the Elderly, edited by James W. Cooper, PharmPhD, and William E. Wade, PharmD (Vol. 12, No. 1, 1997). *"This book is clinician-friendly. . . . Pharmacists caring for elderly patients will find the information contained within the book useful in optimizing their provision of pharmaceutical care." (Jeffrey C. Delafuente, MS, FCCP, Professor and Associate Chairman, Department of Pharmacy Practice, College of Pharmacy, University of Florida, Gainesville)*

Geriatric Drug Therapy Interventions, edited by James W. Cooper, PharmPhD (Vol. 11, No. 4, 1997). *Explores how interventions in geriatric drug therapy can improve drug adherence and reduce adverse drug reactions as well as contribute to improved disease state management in older patients.*

Urinary Incontinence in the Elderly: Pharmacotherapy Treatment, edited by James W. Cooper, PharmPhD (Vol. 11, No. 3, 1997). *"A very useful reference for clinicians working with incontinent patients in various settings." (Susan W. Miller, PharmD, FASCP, Professor, Department of Pharmacy Practice, Mercer University Southern School of Pharmacy, Atlanta, Georgia)*

Antivirals in the Elderly, edited by James W. Cooper, PharmPhD (Vol. 10, No. 2, 1997). *"The chapters are focused and full of useful detail. . . . Will help family physicians and those dealing with the elderly." (Canadian Family Physician)*

Antiinfectives in the Elderly, edited by James W. Cooper, PharmPhD (Vol. 10, No. 1, 1997). *"A useful resource for those treating common infections in the geriatric patient. This text should be extensively utilized." (Keith D. Campagna, PharmD, BCPS, Associate Professor, School of Pharmacy, Auburn University; Clinical Associate Professor, School of Medicine, University of Alabama at Birmingham)*

Geriopharmacotherapy in Home Health Care: New Frontiers in Pharmaceutical Care, edited by Steven R. Moore, RPh, MPH (Vol. 7, No. 3, 1993). *The contributing authors examine common problems and perceptions across states, both in medications and other programmatic concerns for the elderly.*

Diabetes Mellitus in the Elderly

James W. Cooper
Editor

Diabetes Mellitus in the Elderly has been co-published simultaneously as *Journal of Geriatric Drug Therapy*, Volume 12, Number 2 1999.

Pharmaceutical Products Press
An Imprint of
The Haworth Press, Inc.
New York • London

Published by

Pharmaceutical Products Press®, 10 Alice Street, Binghamton, NY 13904-1580 USA

Pharmaceutical Products Press® is an imprint of The Haworth Press, Inc., 10 Alice Street, Binghamton, NY 13904-1580 USA.

Diabetes Mellitus in the Elderly has been co-published simultaneously as *Journal of Geriatric Drug Therapy*™, Volume 12, Number 2 1999.

© 1999 by The Haworth Press, Inc. All rights reserved. No part of this work may be reproduced or utilized in any form or by any means, electronic or mechanical, including photocopying, microfilm and recording, or by any information storage and retrieval system, without permission in writing from the publisher. Printed in the United States of America.

The development, preparation, and publication of this work has been undertaken with great care. However, the publisher, employees, editors, and agents of The Haworth Press and all imprints of The Haworth Press, Inc., including The Haworth Medical Press® and Pharmaceutical Products Press®, are not responsible for any errors contained herein or for consequences that may ensue from use of materials or information contained in this work. Opinions expressed by the author(s) are not necessarily those of The Haworth Press, Inc.

Cover design by Thomas J. Mayshock Jr.

Library of Congress Cataloging-in-Publication Data

Diabetes mellitus in the elderly / James W. Cooper, editor.
    p. cm.
    "Diabetes mellitus in the elderly has been co-published simultaneously as Journal of geriatric drug therapy, volume 12, number 2, 1999."
    Includes bibliographical references and index.
    ISBN 0-7890-0682-0 (alk. paper)
    1. Diabetes in old age.   I. Cooper, James W, 1944- .
RC660.75.D53 1999
618.97'6462–dc21   99-12436
                                                                                     CIP

INDEXING & ABSTRACTING

Contributions to this publication are selectively indexed or abstracted in print, electronic, online, or CD-ROM version(s) of the reference tools and information services listed below. This list is current as of the copyright date of this publication. See the end of this section for additional notes.

- *Abstracts in Social Gerontology: Current Literature on Aging*
- *Adis International Ltd*
- *AgeInfo CD-Rom*
- *AgeLine Database*
- *Applied Social Sciences Index & Abstracts (ASSIA) (Online: ASSI via Data-Star) (CDRom: ASSIA Plus)*
- *Biosciences Information Service of Biological Abstracts (BIOSIS)*
- *Brown University Geriatric Research Application Digest "Abstracts Section"*
- *Brown University Long-Term Care Quality Letter "Abstracts Section"*
- *BUBL Information Service: An Internet-based Information Service for the UK higher education community*
- *Cambridge Scientific Abstracts*
- *CNPIEC Reference Guide: Chinese National Directory of Foreign Periodicals*
- *Current Awareness in Biological Sciences (C.A.B.S.)*
- *EMBASE/Excerpta Medica Secondary Publishing Division*
- *Family Studies Database (online and CD/ROM)*
- *Human Resources Abstracts (HRA)*
- *Index to Periodical Articles Related to Law*
- *International Pharmaceutical Abstracts*

(continued)

- *New Literature on Old Age*
- *Psychological Abstracts (PsycINFO)*
- *Referativnyi Zhurnal (Abstracts Journal of the All-Russian Institute of Scientific and Technical Information)*
- *Social Planning/Policy & Development Abstracts (SOPODA)*
- *Social Work Abstracts*
- *Sociological Abstracts (SA)*

Special Bibliographic Notes related to special journal issues (separates) and indexing/abstracting:

- indexing/abstracting services in this list will also cover material in any "separate" that is co-published simultaneously with Haworth's special thematic journal issue or DocuSerial. Indexing/abstracting usually covers material at the article/chapter level.
- monographic co-editions are intended for either non-subscribers or libraries which intend to purchase a second copy for their circulating collections.
- monographic co-editions are reported to all jobbers/wholesalers/approval plans. The source journal is listed as the "series" to assist the prevention of duplicate purchasing in the same manner utilized for books-in-series.
- to facilitate user/access services all indexing/abstracting services are encouraged to utilize the co-indexing entry note indicated at the bottom of the first page of each article/chapter/contribution.
- this is intended to assist a library user of any reference tool (whether print, electronic, online, or CD-ROM) to locate the monographic version if the library has purchased this version but not a subscription to the source journal.
- individual articles/chapters in any Haworth publication are also available through the Haworth Document Delivery Service (HDDS).

Diabetes Mellitus in the Elderly

CONTENTS

Introduction *James W. Cooper*	1
SILVER THREADS	
Diabetes, Self-Care and Divine Intervention *James W. Cooper*	3
ARTICLES	
The Pathophysiology of Diabetes in Aging *Nir Barzilai* *Meredith Hawkins*	5
The Management of Type 2 Diabetes in the Elderly Patient *John R. White, Jr.* *R. K. Campbell*	21
Acarbose: 'The European Experience' *Maria M. Byrne* *Moria A. Burhorn* *Burkhard Göke*	47
Insulin Use in the Elderly *Stephen N. Davis* *Jeri B. Brown*	61
Index	83

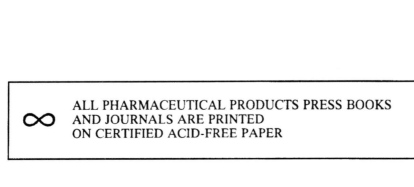

ALL PHARMACEUTICAL PRODUCTS PRESS BOOKS
AND JOURNALS ARE PRINTED
ON CERTIFIED ACID-FREE PAPER

ABOUT THE EDITOR

James W. Cooper, PharmPhD, BCPS, FASCP, FASHP, is Professor of Pharmacy Practice at the University of Georgia College of Pharmacy in Athens, Georgia, and Assistant Clinical Professor of Family Medicine at the Medical College of Georgia. He is the author or editor of 30 books and monographs, 10 book chapters, and over 400 research and professional publications. He teaches, practices, and conducts research in consultant pharmacy with geriatric patients in ambulatory and long-term care settings. The editor of the *Journal of Geriatric Drug Therapy*, Dr. Cooper is a board-certified pharmacotherapy specialist and a Fellow of the American Societies of Consultant Pharmacists and Health Systems Pharmacists. In 1981, he was a special advisor to the White House Conference on Aging. He is the recipient of numerous national awards for his work, and has received more than a million dollars in funding to support his research and service.

Introduction

This volume is devoted to diabetes mellitus diagnosis and treatment in the elderly. This book updates an earlier devoted issue of the *Journal of Geriatric Drug Therapy* (Vol. 5, No. 1, 1990). Silver Threads addresses the frustration and futility health care providers and ministers sometimes experience in attempting to help diabetic patients care for themselves. Barzilai and Hawkins present a review of the pathophysiology of diabetes in aging. A better understanding of the multiorgan involvement in the pathophysiology of diabetes is gained from this paper. The recently revised American Diabetes Association (ADA) guidelines to include impaired glucose tolerance and a lower level of fasting plasma glucose for the diagnosis of diabetes underscore the need for earlier detection of diabetes in the older adult and the relationship of diabetes with obesity.[1] White and Campbell provide the latest information on the management of type 2 diabetes in the elderly patient, with an emphasis on oral agents. With four new oral agents, acarbose, metformin, repaglinide and troglitazone for type 2 disease, and the ADA revised guidelines[1] that suggest the trial of all oral agents before resorting to insulin, the authors offer practical insight to the rational use of these agents. Byrne, Burhorn and Göke discuss the European experience with acarbose. Davis and Brown emphasize the safe use of insulin in the elderly with either type 1 or type 2 diabetes. This review discusses the comorbidities, nutritional issues, pharmacodynamics and physiologic effects of various insulin preparations and regimens.

James W. Cooper

REFERENCE

1. The Expert Committee on the Diagnosis and Classification of Diabetes Mellitus Report. ADA Guidelines. Diabetes Care 1997; 20:1183-1197.

[Haworth indexing entry note]: "Introduction." Cooper, James W. Published in *Diabetes Mellitus in the Elderly* (ed: James W. Cooper) Pharmaceutical Products Press, an imprint of The Haworth Press, Inc., 1999, p. 1. Single or multiple copies of this article are available for a fee from The Haworth Document Delivery Service [1-800-342-9678, 9:00 a.m. - 5:00 p.m. (EST). E-mail address: getinfo@haworthpress inc.com].

SILVER THREADS

Diabetes, Self-Care and Divine Intervention

A minister related to me the following story: I once knew a diabetic woman with a heart condition, who totally ignored her doctor's advice. She was very heavy, but neither watched her diet nor exercised. She smoked, drank and ate sugar-filled desserts. She was careless about taking her medications. Not surprisingly, I got a call one day that she'd collapsed and was in the emergency room. I remember her wide and frightened eyes peering at me over the oxygen mask as she said, "I know God will get me through this!"

She died.

God had tried to help her avoid the hospital altogether, but she had refused to consider the possibility that God might speak to her through her doctor's voice. She would accept God's help only if God refrained from working through human agency, including her own heart, mind, and will. After tying God's hands, she tried to foist responsibility for her condition off on God.

I am still angry at her. Lord help me recognize your voice.

James W. Cooper
Editor

ARTICLES

The Pathophysiology of Diabetes in Aging

Nir Barzilai
Meredith Hawkins

SUMMARY. Glucose levels are tightly regulated by the combined function of the *muscle* to dispose of postprandial glucose, the *liver* to provide for fasting glucose production, and the *β-cells* of the pancreas to regulate both by secreting appropriate amounts of insulin. Aging is known to be an insulin resistant state because hyperinsulinemia is required to maintain fasting and postprandial glucose levels in the normal

---

Nir Barzilai, MD, is affiliated with the Department of Medicine, the Diabetes Research and Training Center, and the Divisions of Geriatrics and Endocrinology, Albert Einstein College of Medicine, Bronx, NY. Dr. Barzilai is a recipient of the Paul Beeson Physician Faculty Scholar in Aging Award. Meredith Hawkins, MD, is affiliated with Department of Medicine, the Diabetes Research and Training Center, and the Division of Endocrinology, Albert Einstein College of Medicine, Bronx, NY.

Address correspondence to: Nir Barzilai, MD, Division of Endocrinology, Department of Medicine, Albert Einstein College of Medicine, Befer Building #701, 1300 Morris Park Avenue, Bronx, NY 10461 (E-mail: barzilai@aecom.yu.edu).

Dr. Barzilai is supported by grants from the National Institutes of Health (KO8-AG00639, R29-AG15003). Dr. Hawkins is the recipient of a fellowship from the Medical Research Council of Canada.

[Haworth co-indexing entry note]: "The Pathophysiology of Diabetes in Aging." Barzilai, Nir, and Meredith Hawkins. Co-published simultaneously in *Journal of Geriatric Drug Therapy* (Pharmaceutical Products Press, an imprint of The Haworth Press, Inc.) Vol. 12, No. 2, 1999, pp. 5-20; and: *Diabetes Mellitus in the Elderly* (ed: James W. Cooper) Pharmaceutical Products Press, an imprint of The Haworth Press, Inc., 1999, pp. 5-20. Single or multiple copies of this article are available for a fee from The Haworth Document Delivery Service [1-800-342-9678, 9:00 a.m. - 5:00 p.m. (EST). E-mail address: getinfo@haworthpressinc.com].

© 1999 by The Haworth Press, Inc. All rights reserved.

range. Insulin resistance is associated with increases in total and visceral fat mass which are typical of aging. Insulin resistance is also associated with risk factors for accelerated atherosclerosis and coronary artery disease, hypertension and hyperlipidemia. The liver maintains normal glucose levels postprandially and during fasting, but with aging more insulin is required to appropriately regulate hepatic glucose production and avoid hyperglycemia. While β-cell insulin secretion may compensate for the resistance to insulin action of the muscle and liver, elderly subjects with and without obesity may fail to respond by secreting adequate amounts of insulin, and will develop diabetes mellitus. The onset of frank diabetes mellitus is accompanied by further deterioration in muscle, liver, and β-cell function, a phenomenon referred to as "glucose toxicity." Understanding the multi-organ pathophysiology of diabetes in the elderly is clinically relevant, because present and future pharmacologic therapies aim to reverse specific organ defects, and often act synergistically to decrease hyperglycemia. *[Article copies available for a fee from The Haworth Document Delivery Service: 1-800-342-9678. E-mail address: getinfo@haworthpressinc.com]*

KEYWORDS: diabetes mellitus, aging, pathophysiology, visceral fat, insulin resistance

TYPE 2 DIABETES MELLITUS (DM)

Type 2 diabetes mellitus (Type 2 DM, previously referred to as non-insulin-dependent diabetes mellitus) is a heterogeneous group of disorders in which hyperglycemia results from both an impaired insulin secretory response to glucose and decreased insulin action (insulin resistance) at the level of the muscle and the liver.[1] The diagnosis of diabetes was previously made in individuals with fasting plasma glucose levels of >140 mg/dL. A recent American Diabetes Association Consensus states that fasting plasma glucose levels of > 125 mg/dL are sufficient for the diagnosis.[2] Type 2 DM is usually diagnosed in patients over forty years of age, but can also develop in children and adolescents. As will be further discussed below, upper body obesity (abdominal/visceral adiposity) is a risk factor associated with development of Type 2 DM. Obese patients with Type 2 DM may actually regain normal glucose levels after a period of weight reduction.

Patients with Type 2 DM only rarely develop diabetic ketoacidosis (DKA) in which insulin therapy is not mandatory for survival, as opposed to patients with Type 1 DM (previously referred to as insulin-dependent diabetes mellitus). While most patients are treated with a combination of diet, exercise and oral agents, some patients with Type 2 DM will intermittently or persistently require insulin to control symptomatic hyperglycemia and prevent non-ketotic hyperglycemic-hyperosmolar coma.

The concordance rate for Type 2 DM in monozygotic twins is > 90%, and genetic factors appear to be the major determinants of its development. Type 2 DM with an autosomal dominant inheritance pattern has been found in successive generations of some families, frequently in asymptomatic, nonobese, young adolescents.

Many families with maturity-onset diabetes of the young (MODY) have a mutation in the gene for glucokinase,[3] the major glucose phosphorylating enzyme in the β-cell and liver, and impairments in both insulin secretion and in hepatic glucose regulation have been demonstrated in these patients.[4]

Other than this rare example, single gene mutations have not been demonstrated as a cause for diabetes, and it is commonly believed now that Type 2 DM is a polygenic disease, which is manifested in the presence of plentiful availability of energy stores. No association between Type 2 DM and specific HLA phenotypes or islet-cell antibodies (ICA) has been demonstrated (an exception is a subset of nonobese adults with detectable ICA who carry one of the HLA phenotypes and who may eventually develop Type 2 DM).

The pancreatic islets in Type 2 DM are not consistently altered, and normal β-cell mass appears to be preserved in most patients. Pancreatic islet amyloid, resulting from a deposition of amylin, is found in a high percentage of Type 2 DM patients at autopsy, but its relationship to the pathogenesis of Type 2 DM is not well established.

DIABETES AND IMPAIRED GLUCOSE TOLERANCE WITH AGING

The incidence of Type 2 DM, as defined by the previous WHO criteria,[1] is increased ~3 fold in the elderly population and is often undiagnosed.[5-9] The prevalence of diabetes mellitus is ~10-20% in both men and women in the 6-7th decade as compared with a prevalence of ~3-5% at the 4-5th decade of life.[5-9] These figures are expected to increase with the new criteria for diagnosing DM (vide supra), i.e., fasting plasma glucose >125 mg/dL. In addition, many elderly subjects fall into the category of impaired glucose tolerance (IGT; fasting plasma glucose 110-125 mg/dL) bringing the overall prevalence of abnormal glucose metabolism with aging to greater than 30%.[10] A common metabolic alteration associated with aging is a marked increase in stimulated plasma glucose levels,[5-9] while fasting glucose levels increase only mildly with aging. However, it has not been determined unequivocally whether any of these alterations in glucose metabolism are due to aging per se or to concomitant diseases and their treatments, to the decrease in physical activity, or most importantly, to the typical changes in body composition that occur with aging.

The most common alterations related to impaired glucose metabolism in aging are progressive rises in fasting and post-prandial plasma insulin levels,[5-9] suggesting that aging is an insulin resistant state. Recently, interest has been focused on the common epidemiological association of insulin resistance with obesity, dyslipidemia, hypertension, hyperuricemia, dysfibrinolysis and IGT or diabetes mellitus, commonly known as the *syndrome of insulin resistance* (or syndrome "X").[11-12] This syndrome is also associated with a markedly increased risk of developing coronary heart disease and other macrovascular disease, and is therefore linked to a decreased life expectancy.[13-14] Interestingly, an increased fasting plasma insulin level, which may serve as a clinical marker for insulin resistance, is an isolated risk factor for the development of coronary heart disease.[15] Moreover, hyperinsulinemia and insulin resistance are often considered to be the earliest events leading to diabetes mellitus. Since the most common association with the syndrome of insulin resistance is obesity, we will focus on changes in body composition with aging as the significant event initiating impaired insulin action and hyperinsulinemia.

THE TYPICAL ALTERATIONS IN BODY COMPOSITION IN HUMAN AGING

Data from the National Health and Nutrition Examination Survey (NHANES) III indicate that approximately half of the US population over age 50 is obese.[16] Obesity is due to the increased proportion of fat to muscle mass, which typically occurs between the third and seventh decade of life; thereafter, fat mass may increase further, remain unchanged, or decrease.[17-19] Despite a modest age-related decrease in lean body mass, the ratio of fat mass over lean body mass increases throughout human life.[20-21]

Increased visceral fat, in particular, is a common and typical change in body composition with aging.[22-24] This increase in visceral fat can be detected surgically (increased omental fat), or by computerized axial tomography (CT) scan or magnetic resonance imaging (MRI). However, the most common epidemiological tools are observation of a pattern of upper body obesity, increased waist circumference, and increased waist-to-hip ratio (WHR).[17] CT studies in men and women of all ages revealed a significant inverse correlation between the ratio of subcutaneous to visceral fat tissue with age, and an increased accumulation of fat between the muscles of the abdominal wall in older subjects.[23-24] Of most importance, the WHR in older men and women increases independently from body mass index (BMI)[22] (please see Figure 1). Thus, older men and women show an increase in visceral fat, although BMI may actually be normal and lean body mass may be decreased.

The increased visceral fat of aging has associations with the endocrine changes of both male and female menopause, neuroendocrine failure, and lifestyle changes, and may be genetically determined by a major autosomal recessive locus. Moreover, testosterone and estrogen replacement, and increased exercise have shown to decrease visceral fat in older subjects.[25-29]

Increased visceral fat was an independent risk factor for the development of stroke and coronary artery disease in a long term prospective study beginning at late middle-age.[30] WHR was specifically found to be a better marker than BMI for risk of death in older women.[31] Increased visceral fat was also an important risk factor for death from coronary artery disease in postmenopausal women.[32] *Thus, visceral fat has multi-factorial associations with aging, and may determine mortality in humans from atherosclerotic mechanisms.*

Finally, numerous studies in subjects of all ages have demonstrated that increased visceral fat is associated with increases in fasting and post-prandial plasma insulin levels and IGT.[32-34] Furthermore, of a cohort of middle-aged men followed for ~20 years, those who maintained their WHR in the lower decile had no increased risk for devel-

FIGURE 1. Change in Body Composition with Aging (Waist to hip ratio (bottom panel) and BMI (kg/m^2) (upper panel) according to age. While BMI is often similar to young controls, abdominal obesity is typically and markedly increased with aging. Adapted from Shimota et al. Ref. 21.)

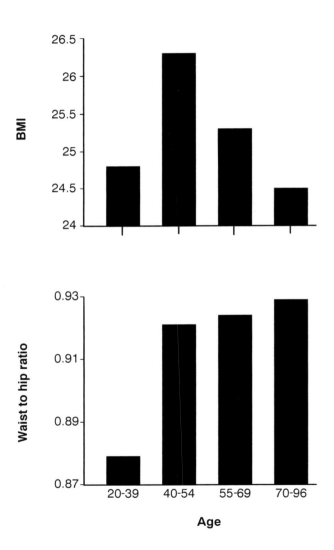

oping diabetes mellitus, while subjects with the highest decile had ~20 fold increased relative risk for developing diabetes mellitus.[29] *Thus, in weight-matched subjects, hyperinsulinemia, IGT and diabetes mellitus occur significantly more often in the presence of increased visceral fat mass than with increased total fat mass.*

MECHANISMS BY WHICH GLUCOSE HOMEOSTASIS IS IMPAIRED IN HUMAN AGING: THE PATHWAY LEADING TO DIABETES

Impaired regulation in glucose homeostasis with aging will be reviewed from the perspective of the *muscle* (the major site for post-meal glucose disposal), the *liver* (the major site for fasting glucose production), and the β-cell. As depicted in Figure 1, increases in visceral/abdominal fat with aging may determine the pathophysiologic changes in these organs leading to diabetes, and ultimately to atherosclerotic disease.

Decreases in Insulin-Mediated Glucose Uptake with Aging

Skeletal muscle accounts for the majority of the body's insulin-mediated glucose uptake. Thus, the typical increases in plasma insulin levels with aging suggest peripheral (skeletal muscle) insulin resistance.[35-35a] Many investigators have demonstrated an age-related decline in whole-body glucose utilization (wbgu) when expressed in terms of total body mass. This decline in wbgu which does not correct for large amounts of relatively inactive adipose tissue and results in an underestimation of glucose uptake by lean body mass.[36-38] Interestingly, decline in wbgu is most rapid between the ages of 20 to 30,[36] which may reflect that population's greater exercise capacity (higher VO_2 max) and lower fat mass. Indeed, fat mass was inversely correlated with insulin-mediated glucose uptake, independent of age,[39-40] and the effects of body composition on insulin sensitivity, glucose utilization in lean older individuals was similar to that in weight-matched young individuals. A recent multi-center study demonstrated that insulin resistance did not decrease with aging, and most of the variation in insulin action could be attributed to increase in body weight and fat mass, as demonstrated in Figure 2.[41]

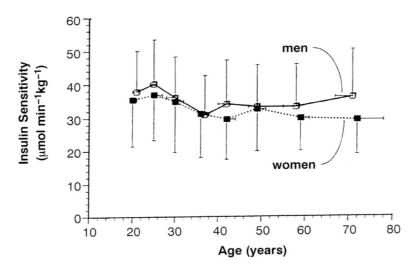

FIGURE 2. Insulin Action in Aging (Insulin sensitivity was studied by hyperinsulinemic euglycemic clamp techniques in 1,146 men and women by decade of age. While insulin sensitivity does not decline with aging, the large standard deviations (vertical bars are ± 1 SD) are due to differences in BMI. Thus, BMI more than aging determines insulin action. Adapted from Ferrannini et al. Ref. 40.)

Furthermore, increased visceral fat has been associated with marked decreases in skeletal muscle insulin sensitivity and responsiveness in BMI-matched women with upper vs. lower body obesity.[42] Indeed, visceral fat accounted for 79% of the variability in insulin sensitivity in another heterogeneous population of women.[43] Interestingly, in a relatively lean population of black Americans with diabetes mellitus, peripheral insulin resistance was demonstrated in the subjects with increased visceral fat as determined by CT scans.[44] Finally, insulin sensitivity in healthy older men was affected in decreasing order by visceral fat, obesity, and VO_2 max.[45]

Once glucose is phosphorylated it is directed intracellularly into storage (mainly as glycogen) or oxidation (via the glycolytic pathway). Measurements of the proportion of glucose directed to storage vs. oxidation both in obesity and in diabetes mellitus have generally demonstrated a defect in glucose storage rather than oxidation.[46] However, this relationship has not yet been effectively studied in human aging. Cloning of proteins that determine insulin action (such as the insulin receptor, part of its signaling pathway, and the glucose transporter system) enabled investigators to rule out genetic mutations in these proteins as a common cause for insulin resistance or DM. *In vitro* studies failed to demonstrate any consistent alteration in either insulin binding to monocytes[47] or in the insulin-sensitive glucose transporter system (GLUT4) in adipose cells[48] with aging. Finally, while the content of GLUT4 (the insulin dependent glucose transporter) was shown to be decreased in some muscle types with aging,[49] it is the translocation of GLUT4 from low density microsomes to the plasma membrane that is believed to be more relevant to insulin-mediated glucose transport in animal and human muscle.

Thus, while fat mass has a profound effect on insulin sensitivity in aging, increased visceral fat appears to be an even more potent determinant of insulin resistance. Hence, the decline in insulin-mediated peripheral glucose uptake may be associated with factors other than aging, such as changes in body composition and in the relative contributions of different metabolic pathways.

As will be discussed in other chapters, agents which enhance peripheral insulin action are being developed for use in the treatment of Type 2 DM. The first drug from the family of thiazolidinediones, troglitazone, is now available on the US market.

Progressive Decreases in the Ability of Insulin and Glucose to Regulate Hepatic Glucose Fluxes in Aging

While hepatic glucose production (HGP) is unchanged in aging,[36-38,50-52] it has often been measured in the presence of increased plasma insulin and glucose levels. These findings suggest that the aging liver is resistant to the inhibitory effects of insulin and glucose on HGP.

Hepatic glucose production during hyperinsulinemia showed a ~4 fold increase in the highest vs. the lowest tertile of obese subjects with increased visceral fat.[53] Similarly, approximately ~2-3 fold increases in portal insulin levels were needed to suppress HGP in women with upper body obesity as compared with lower body obesity.[54] Hepatic insulin sensitivity seems to decrease even more than peripheral insulin sensitivity with increased visceral fat, suggesting a selective effect of visceral fat on the liver. Thus, increased visceral fat is better correlated than obesity with decreased hepatic insulin action.

Biguanide drugs, e.g., metformin, are proven to be effective in decreasing hepatic glucose production through unknown mechanisms.

Impairments in Insulin Secretion with Aging

Since a large proportion of insulin-resistant subjects develop diabetes mellitus and IGT with aging, this has been hypothesized to be due to a failure by the β-cell to secrete adequate amounts of insulin. An absolute decrease in insulin secretory rates has been demonstrated in some, but not all, studies of human aging.[55-57] Decreased insulin secretion may be due, in part, to genetic background, decreased physical activity, and changes in body composition, none of which has been adequately controlled in these studies. Peripheral insulin resistance is associated with increased insulin secretion in young populations with normal glucose tolerance.[58-59] In elderly subjects, insulin secretion was decreased compared with young subjects with similar insulin responsiveness, suggesting a negative effect of age on β-cell function.[60]

In insulin-resistant obese subjects, the development of IGT was determined by decreased β-cell function and not by any further reduction in insulin sensitivity, suggesting that increased fat mass may determine the eventual decrease in insulin secretion.[61] In addition, a recent study demonstrated that while a significant subset of lean elderly subjects developed diabetes mellitus with no significant insulin resistance, those subjects had decreased insulin responses to hyperglycemia, suggesting a defect in β-cell secretion with aging.[62] Thus, in those obese patients who develop Type 2 DM, the initial compensatory increases in insulin secretion have eventually failed. However, as depicted in Figure 3, with advanced age alone insulin secretion may fail with no previous evidence for insulin resistance and prolonged insulin hypersecretion. *Therefore, the combination of aging, changes in body composition, and peripheral insulin resistance may determine the ultimate failure in insulin secretion. Such*

FIGURE 3. Depiction of Insulin Secretion in Aging (Insulin secretion may decrease in aging, leading to diabetes in the elderly lean population. More commonly, obesity in middle age leads to insulin resistance and compensatory hyperinsulinemia. This may lead to a more rapid decrease in the capacity of insulin secretion with age, and early onset of diabetes. The broken line depicts the minimal insulin levels that are needed to prevent ketoacidosis, which is a relatively late occurrence in all forms of Type 2 diabetes.)

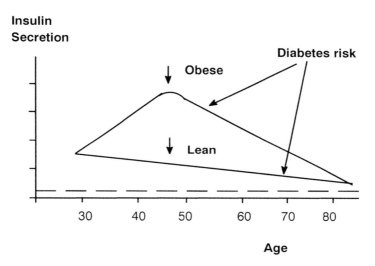

failure in insulin secretion may be responsive to treatment with sulfonylurea drugs.

Insulin secretion was often correlated with obesity and insulin resistance as mentioned above, but correlations with visceral fat have not been adequately established. In obese, non-diabetic girls, basal insulin secretion and stimulated insulin secretion were highly correlated with visceral fat (viewed by MRI), but not with subcutaneous fat.[59] While young subjects adequately compensate for the development of insulin resistance by increasing insulin secretion, the increased incidence of diabetes mellitus with increased age and visceral fat, suggests a greater frequency of β-cell failure.

PREVENTION OF TYPE 2 DM

Several lines of evidence indicate that Type 2 DM may be prevented or delayed by lifestyle and pharmacologic interventions in high risk populations (relatives of Type 2 DM, women after gestational diabetes, etc.). The National Institutes of Health recently funded a large multicentral trial to examine this hypothesis. The arms of the study include changes in diet, physical activity, and the antihyperglycemic agents metformin and troglitazone. Hopefully this effort may not only improve body weight and body composition, but might also increase insulin action on the liver and muscle, thus preventing hyperglycemia.

FRANK DIABETES AND "GLUCOSE TOXICITY"

The onset of frank diabetes mellitus and hyperglycemia occurs following the gradual age-related progression of defects in skeletal muscle glucose uptake, hepatic insulin action and pancreatic insulin secretion. Chronic increases in plasma glucose levels are then associated with further increases in insulin resistance and hepatic glucose production, and a further deterioration in the capacity of the β-cell to secrete insulin. This "snow ball" phenomenon, commonly known as "*Glucose Toxicity,*" refers to the deleterious effects of prolonged hyperglycemia per se on both insulin action and secretion, suggesting that hyperglycemia is not only a consequence but also a cause of further impairments in glucose tolerance in the diabetic patient.[63] Lowering glucose levels by many means (such as diet, exercise, sulfonylurea, or insulin), has been shown to result in improved insulin sensitivity and insulin secretion, and in decreased hepatic glucose production.

In summary, human aging is associated with significant alterations in glucose metabolism and insulin action, together with a 3-fold higher prevalence of diabetes mellitus and ICT. This paper outlined the contributions of age-related changes to the roles of the three major organs involved in glucose regulation, i.e., the skeletal muscle, the liver and the pancreatic β-cell. One of the most important etiologic factors associated with all of the above defects is the typical change in body composition which occurs with aging, i.e., the gradual development of visceral adiposity (Figure 4).

FIGURE 4. Hypothetical Model of How Increased Visceral Fat with Aging Leads to Diabetes and Atherosclerotic Vascular Diseases

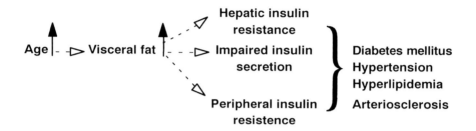

REFERENCES

1. Rifkin H, Porte D. Jr. in Diabetes Mellitus, New York: Elsevier Science Publishing Co., Inc, 1990, pp. 346-464.
2. The Expert Committee on the Diagnosis and Classification of Diabetes Mellitus. Diabetes Care 1998, 21:(Suppl. 1) S5-S31.
3. Froguel PH, Vaxillaire M, Sun F, et al. Close link of glucokinase locus on chromosome 7p to early-onset non-insulin-dependent diabetes mellitus. Nature 1992; 356:162-164.
4. Velho G, Clement K, Pueyo ME, et al. Hepatic insulin resistance in MODY patients carrying mutations in the glucokinase gene. Diabetes 1994; 42: 244A.
5. Reaven GM, Reaven EP. Age, glucose intolerance, and non-insulin-dependent diabetes mellitus. J Am Geriatr Soc. 1985; 33:286-290.
6. Fraze E, Chiou M, Chen Y, Reaven GM. Age related changes in postprandial plasma glucose, insulin, and FFA concentrations in non-diabetic individuals. J Am Geriatr Soc. 1987; 35:224-228.
7. Davidson MB. The effect of aging on carbohydrate metabolism: a review of the english literature and a practical approach to the diagnosis of diabetes mellitus in the elderly. Metabolism 1979; 28:688-705.
8. Goldberg AP, Andres R, Bierman EL. Diabetes mellitus in the elderly. In: Andres R, Bierman EL, Hazzard WR, eds. Principles of geriatric medicine. New York: McGraw-Hill 1985, p. 311.
9. Weingard DL, Sinsheimer P, Barrett-Connor EL, McPhillip JB. Community-based study of prevalence of NIDDM in older adults. Diabetes Care 1990; 13 (Suppl. 2): 3-8.
10. Harris MI. Impaired glucose tolerance in the U.S. population. Diabetes Care 1989; 12(7):464-74.

11. Modan M, Halkin H, Lusky A, et al. Hyperinsulinemia a link between hypertension obesity and glucose tolerance. J Clin Invest. 1985; 75:809-817.

12. Reaven GM. (Banting) Role of insulin resistance in human disease. Diabetes 1988; 37:1595-1607.

13. Zavaroni I, Bonora E, Pagliara M, et al. Risk factors for coronary artery disease in healthy persons with hyperinsulinemia and normal glucose tolerance. N Engl J Med 1989; 320:702-706.

14. DeFronzo RA, Ferrannini E. Insulin resistance: a multifaceted syndrome responsible for NIDDM, obesity, hypertension, dyslipidemia, and atherosclerosis cardiovascular disease. Diabetes Care 1991; 14:173-194.

15. Despres JP, Lamarche B, Mauriege P, et al. Hyperinsulinemia as an independent risk factor for ischemic heart disease. N Engl J Med 1996; 334:952-7.

16. Kuczmarski RJ, Flegal K, Campbell S, Johnson CL. Increasing prevalence of overweight among US adults. The National Health and Nutrition Examination Surveys, 1960 to 1991 (NHANES III). JAMA 1994; 272:238-239.

17. Andres R. Principles of geriatric medicine. New York: McGraw-Hill, 1985, p. 311.

18. Norris AH, Lundy T, Shock NW. Trends in indices of body composition in men between ages 30-70 years. Annals of the New York Academy of Science 1963; 110:623-639.

19. Cohn SH, Vartsky D, Yasumura S, et al. Compartmental body composition based on total-body nitrogen, potassium and calcium. Am J Physiol. 1980; 239:E524-530.

20. Cohn SH, Vartsky D, Yasumura S, et al. Compartmental body composition based on total-body nitrogen, potassium and calcium. Am J Physiol. 1980; 239:E524-530.

21. Forbes GB, Reina JC. Adult lean body mass declines with age: some longitudinal observations. Metabolism 1970; 19:653-663.

22. Shimokata H, Tobin JD, Muller DC, et al. Studies in the distribution of body fat: 1. effects of age sex, and obesity. J Gerontol. 1989; 44:M66-M73.

23. Enzi G, Gasparo M, Binodetti PR, et al. Subcutaneous and visceral fat distribution according to sex, age, and overweight, evaluated by computed tomography. Am J Clin Nutr. 1986; 44:739-746.

24. Borkan GA, Hults DE, Gerzof SG, et al. Age changes in body composition revealed by computed tomography. J Gerontology 1983; 38: 673-677.

25. Marin P, Holmang S, Jonsson L, et al. The effects of testosterone treatment on body composition and metabolism in middle-aged obese men. Int J Obes Relat Metab Disord. 1992; 16:991-7.

26. Kritz-Silverstein D, Barrett-Connor E. Long-term postmenopausal hormone use, obesity, and fat distribution in older women. JAMA 1996; 275:46-9.

27. Kohrt WM, Obert KA, Holloszy JO. Exercise training improves fat distribution patterns in 60- to 70-year-old men and women. J Gerontol. 1992; 47:M99-105.

28. Marin P, Krotkiewski M, Bjorntorp P. Androgen treatment of middle-aged, obese men: effects on metabolism muscle and adipose tissues. Eur J Med. 1992; 1:329-36.

29. Bouchard C, Rice T, Lemieux S, et al. Major gene for abdominal visceral fat area in the Quebec Family Study. Int J Obes Relat Metab Disord. 1996; 20:420-7.

30. Larson B. Regional obesity as a health hazard in men-prospective studies. Acta Med. Scan. Suppl. 1992; 723:45-51.

31. Folsom AR, Kaye SA, Sellers TA, et al. Body fat distribution and 5-year risk of death in older women. JAMA. 1993; 269:483-7.

32. Prineas RJ, Folsom AR, Kaye SA. Central adiposity and increased risk of coronary artery disease mortality in older women. Epidemiol. 1993; 3:35-41.

33. Colman E, Toth MJ, Katzel LI, et al. Body fatness and waist circumference are independent predictors of the age-associated increase in fasting insulin levels in healthy men and women. Int J Obes Relat Metab Disord. 1995; 19:798-803.

34. Kissebah AH. Insulin resistance in visceral obesity. [Review]. Int J Obes. 1991; 15:109-15.

35. Bjorntorp P. Metabolic implications of body fat distribution. [Review] Diabetes Care 1991; 14:1132-43.

35a. DeFronzo RA. Glucose intolerance and aging: evidence for tissue insensitivity to insulin. Diabetes 1979; 28:1095-1101.

36. Rowe JW, Minaker KL, Pallotta JA, Flier IS. Characterization of the insulin resistance of aging. J Clin Invest. 1983; 71:1581-1587.

37. Meneilly GS, Minaker K, Elahi D, Rowe JW. Insulin action in aging man: evidence for tissue specific differences at low physiological insulin levels. J Gerontol. 1987; 42:196-201.

38. Boden G, Chen X, DeSantis RA, Kendrick Z. Effects of age and body fat on insulin resistance in healthy men. Diabetes Care 1993; 16:728-733.

39. O'Shaughnessy IM, Kasdort GM, Hottman RG, Kalkhott RR. Does aging intensify the insulin resistance of human obesity. J Clin Endocrinol Metab. 1992; 74:1075-1081.

40. Ferrannini E, Vichi S, Beck-Nielsen H, et al. European Group for the Study of Insulin Resistance (EGIR). Insulin action and age. Diabetes 1996; 45:947-953.

41. Peiris AN, Struve MF, Mueller RA, et al. Glucose metabolism in obesity: influence of body fat distribution. J Clin Endocrinol Metab. 1988; 67:760-7.

42. Carey DG, Jenkins AB, Campbell LV, et al. Abdominal fat and insulin resistance in normal and overweight women: Direct measurements reveal a strong relationship in subjects at both low and high risk of NIDDM. Diabetes 1996; 45:633-8.

43. Banerji MA, Chaiken RL, Gordon D, et al. Does intra-abdominal adipose tissue in black men determine whether NIDDM is insulin-resistant or insulin-sensitive? Diabetes 1995; 44:141-6.

44. Coon PJ, Rogus EM, Drinkwater D, et al. Role of body fat distribution in the decline in insulin sensitivity and glucose tolerance with age. J Clin Endocrinol Metab. 1992; 75:1125-32.

45. DeFronzo RA. Pathogenesis of type 2 (non-insulin dependent) diabetes mellitus: a balanced overview. Diabetologia 1992; 35:389-97.

46. Fink RI, Kolterman OG, Griffin J, Olefsky JM. Mechanisms of insulin resistance in aging. J Clin Invest. 1983; 71:1523-1535.

47. Fink RI, Wallas P, Olefsky JM. Effects of aging on glucose mediated glucose disposal and glucose transport. J. Clin. Invest. 1986; 77:2034-2041.

48. Houmard JA, Weidner MD, Dolan PL, et al. Skeletal muscle GLUT4 protein concentration and aging in humans. Diabetes 1995; 44:555-60.

49. Robert II, Cummins JC, Wolfe RR, et al. Quantitative aspect of glucose production and metabolism in healthy elderly subjects. Diabetes 1982; 31:203-211.

50. Jackson RA, Hawa MI, Roshania RD, et al. Influence of aging on hepatic and peripheral glucose metabolism in human. Diabetes 1988; 37:119-129.

51. Barzilai N, Stessman J, Cohen P, et al. Glucoregulatory hormone influence on hepatic glucose production in the elderly. Age 1989; 12:13-17.

52. Carey DG, Jenkins AB, Campbell LV, et al. Abdominal fat and insulin resistance in normal and overweight women: Direct measurements reveal a strong relationship in subjects at both low and high risk of NIDDM. Diabetes 1996; 45:633-8.

53. Peiris AN, Struve MF, Mueller RA, et al. Glucose metabolism in obesity: influence of body fat distribution. I Clin Endocrinol Metab. 1988; 67:760-7.

54. Fink RI, Revers RR, Kolterman OG, Olefsky JM. The metabolic clearance rate of insulin and the feedback inhibition of insulin secretion are altered with aging. Diabetes 1985; 34:275-280.

55. Gumbinar B, Polonsky KS, Beltz WF, et al. Effects of aging on insulin secretion. Diabetes 1989; 38:1549-1556.

56. Bourey RE, Kohrt MW, Kirwan JP, et al. Relationship between glucose tolerance and glucose-stimulated insulin response in 65-year-olds. J Gerontol. 1993; 48:M122-M127.

57. Kahn SE, Prigeon RL, McCulloch DK, et al. Quantification of the relationship between insulin sensitivity and beta-cell function in human subjects. Evidence for a hyperbolic function. Diabetes 1993; 42:1663-72.

58. Caprio S, Hyman LD, Limb C, et al. Central adiposity and its metabolic correlates in obese adolescent girls. Am I Physiol. 1995; 269:E118-26.

59. Chen M, Bergman RN, Pacini G, Porte D Jr. Pathogenesis of age-related glucose intolerance in man: insulin resistance and decreased beta-cell function. J Clin Endocrinol Metab. 1985; 60:13-20.

60. Larsson H, Ahren B. Islet dysfunction in obese women with impaired glucose tolerance. Metabolism 1996; 45:502-9.

61. Meneilly GS, Elliott T, Tessier D, et al. NIDDM in the elderly. Diabetes Care 1996; 19:1320-25.

62. Rossetti L, Giaccari A, DeFronzo R. Glucose toxicity. Diabetes Care 1990; 13(6):610-30.

The Management of Type 2 Diabetes in the Elderly Patient

John R. White, Jr.
R. K. Campbell

SUMMARY. This manuscript evaluates the various available pharmacologic agents and combinations of these agents in the management of type 2 diabetes mellitus in the elderly. Currently the incidence of type 2 diabetes in the elderly is on the rise. While goals of glycemic control may be similar in the elderly to the younger type 2 patients, nuances in the pharmacology of the four groups of oral agents and in insulin products may make some regimens preferable in the elderly population. The positive benefits of euglycemic control in all patients with diabetes has been conclusively demonstrated. Currently in the United States sulfonylureas, alpha-glucosidase inhibitors, biguanides, thiazolidinediones, and various insulin products are available for management of hyperglycemia. When choosing an agent the practitioner must consider efficacy of the medication (blood glucose lowering potential), current and goal blood glucose and glycosylated hemoglobin levels, contraindications, side effects, and effect on other metabolic parameters such as insulin concentration, lipids, and body weight. Additionally in the elderly, remaining life expectancy, presence of diabetic complications, presence of coexisting medical or neuropsychiatric disorders, and the patient's willingness and ability to comply with the

---

John R. White, Jr., PharmD, is Associate Professor and Director, Washington State University/Sacred Heart Medical Center Drug Studies Unit, Washington State University, College of Pharmacy, Pullman, WA. R. K. Campbell, RPh, MBA, CDE, is Associate Dean and Professor, Washington State University, College of Pharmacy.

Address correspondence to: John R. White, Jr., PharmD, 25117 West Coulee Hite Road, Reardan, WA 99029 (E-mail: whitej@wsu.edu).

[Haworth co-indexing entry note]: "The Management of Type 2 Diabetes in the Elderly Patient." White, John R. Jr., and R. K. Campbell. Co-published simultaneously in *Journal of Geriatric Drug Therapy* (Pharmaceutical Products Press, an imprint of The Haworth Press, Inc.) Vol. 12, No. 2, 1999, pp. 21-45; and: *Diabetes Mellitus in the Elderly* (ed: James W. Cooper) Pharmaceutical Products Press, an imprint of The Haworth Press, Inc., 1999, pp. 21-45. Single or multiple copies of this article are available for a fee from The Haworth Document Delivery Service [1-800-342-9678, 9:00 a.m. - 5:00 p.m. (EST). E-mail address: getinfo@haworthpressinc.com].

© 1999 by The Haworth Press, Inc. All rights reserved.

proposed diabetes treatment plan must be considered when developing treatment goals and therapeutic regimens. While cost of medication must be considered, it is incidental (less than 2%) when compared to the overall cost of diabetes. As stated above strict glycemic control reduces complications. Therefore, using the most cost-effective approach, the practitioner should choose the regimen that has the best effect on glycemia, as well as weight, insulin concentration, and plasma lipids, without causing undo side effects. In the long run, this approach will be cost effective with a resultant reduction in morbidity and mortality. *[Article copies available for a fee from The Haworth Document Delivery Service: 1-800-342-9678. E-mail address: getinfo@haworthpressinc.com]*

KEYWORDS: type 2 diabetes mellitus, elderly, sulfonylureas, biguanides, thiazolidinediones, insulin, alpha-glucosidase inhibitors, cost-effective

INTRODUCTION

Diabetes mellitus is a problem which is burgeoning among the elderly. In fact, age is considered a risk factor for the development of type 2 diabetes and the incidence increases with age. Diabetes mellitus is 10 times more prevalent in individuals over 65 years of age than in those between ages 20 and 44.[1] Currently over 10% of individuals over 65 years of age living in the United States have diabetes.[2] It is estimated that the population of those greater than 65 years old will grow from the current 28 million to over 50 million by the year 2020. This suggests that type 2 diabetes mellitus will become an even greater public health problem in the future than it is now.

The Framingham study described the development of diabetes in later life to be associated with hypertension, vascular disease, elevated very-low-density-lipoprotein (VLDL) cholesterol, use of diuretics, and obesity.[3] Syndrome X or the Insulin Resistance Syndrome is a frequently referred to grouping of characteristics observed in patients with type 2 disease and includes hyperinsulinemia, hyperglycemia, hyperlipidemia, hypertension, and obesity. These characteristics are important to consider when treating elderly patients with diabetes because many medical interventions focused at one of the problems may have deleterious effects on the others. Thus management of type 2 diabetes is not simply the reduction of blood glucose levels but includes management of all the risk factors for chronic complications.

Type 2 diabetes in the elderly is a complicated syndrome to manage because patients are particularly prone to heart attacks, strokes, blindness, peripheral vascular problems, neuropathy, and amputations. Treatment of the elderly patient is particularly difficult because of problems with hearing, vision, and memory losses, poor or absent teeth, arthritis, and lack of financial resources.[4] The patient with diabetes is 25 times more likely to become blind, 17 times more likely to develop kidney disease, 20 times more likely to develop gangrene, and 2 times more likely to suffer a stroke or a heart attack than aged matched cohorts without diabetes.[5] However, recent studies have demonstrated a tight correlation between development and progression of most of the chronic complications of diabetes and strict glycemic control.[6]

This manuscript will provide an overview of the data supporting the importance of strict glycemic control, goals of glycemic control, choice of oral agents, insulin regimens and combination oral agent insulin therapy in the elderly patient with type 2 disease.

CONTROL AND COMPLICATIONS IN PATIENTS WITH TYPE 2 DIABETES

The most controversial question in the field of diabetes since the discovery of insulin in 1922 by Banting and Best has been "Does strict glycemic control in patients who have diabetes reduce the incidence and slow down the progression of chronic complications?" In many regards this debate has become even more important as new medications have been made available and as technology allowing for self-monitoring of blood glucose (SMBG) and the routine measurement of glycosylated hemoglobin (HbA1c) has been introduced. Prior to this technology of SMBG, when only urine testing was available, it was virtually impossible for a patient with diabetes to achieve and verify euglycemic control. In the past few years several key trials and studies have dramatically demonstrated the importance of strict glycemic control in both patients with type 1 and type 2 diabetes mellitus; the Diabetes Control and Complications Trial,[6] The Wisconsin Epidemiologic Study,[7] and the Ohkubo Study.[8]

The Diabetes Control and Complication Trial (DCCT) was the first randomized, controlled, prospective study to demonstrate the effectiveness of strict glycemic control in reducing complications in patients with type 1 diabetes mellitus.[6] The DCCT evaluated over 1400 patients with type 1 diabetes. The patients were enrolled into either the primary prevention group (i.e., had no significant complications) or the secondary intervention group (patients who exhibited complications). Patients in the primary prevention group were tested to see if strict control would reduce the appearance of complications while patients in the secondary intervention group tested to see if strict glycemic control would reduce the progression of complications. Patients from the two groups were randomized to receive treatment with either intensive or standard therapy. Intensive treatment (also called "experimental" in the DCCT) was defined as therapy with a subcutaneous insulin infusion pump or a regimen of three or more insulin injections per day. Standard therapy was defined as one or two insulin injections per day of intermediate acting insulin. The results of this study were unequivocally favorable in patients treated with intensive therapy. The difference in effectiveness of the two therapies was determined to be so great during an interim analysis of the data that the study was discontinued one year earlier than planned.

The DCCT demonstrated that intensive treatment (strict glycemic control) was associated with a reduction in retinopathy in the patients studied. Researchers reported a 27% reduction in the initial appearance of retinopathy in those treated with intensive therapy. A reduction of 34-76% in clinically significant retinopathy was reported in those who maintained strict glycemic control. Additionally, a 45% reduction in the progression of mild retinopathy to severe retinopathy was observed in those treated with intensive therapy.[6] Additionally, the study reported a 60% reduction in the development of clinical neuropathy in those patients who maintained strict glycemic control.[6]

The appearance and progression of nephropathy was reduced with strict glycemic control in the patients studied. A 56% reduction in the development of clinical grade proteinuria (macroproteinuria $\geq$ 500 mg/day) in patients being treated with intensive therapy was observed. Also, a 35% reduction in the microalbuminuria, a marker for beginning stages of renal dysfunction, was reported in the strict glycemic control group. These data strongly support the notion that strict glycemic control reduces both the appearance and the progression of renal dysfunction in patients with diabetes. It should be noted however that the strict control group experienced a two to three fold increase in severe hypoglycemic episodes when compared to the standard therapy group.

The findings of the DCCT were the first to unequivocally support the premise that tight glycemic control does improve long term outcome in patients with type 1 diabetes. A summary of the results from the DCCT is provided in Table 1 below.

A long term population study in patients with diabetes, The Wisconsin Epidemiologic Study, demonstrated a strong correlation between glycemic control, microvascular and macrovascular complications, and mortality rates in patients with type 1 and type 2 diabetes mellitus.[7] For example, 10-year mortality rates in the patients with older onset diabetes were 35% in patients in the lower glycemic quartile (HbA1c 5.4-8.5) compared to 58% in the upper quartile (HbA1c 11.5-20.8%). This study, while not interventional in design, provided strong evidence for the utility of strict glycemic control in older patients with type 2 diabetes.

An interventional study in patients with type 2 disease, similar in design to the DCCT was reported by Ohkubo et al. in 1995.[8] As with the DCCT, the results overwhelmingly supported the hypothesis that intensive glycemic control can delay the onset and progression of diabetic retinopathy, nephropathy, and neuropathy in patients with

TABLE 1

DCCT[6]–Intensive Insulin Therapy was associated with a:

35% reduction in the development of *microalbuminuria*
56% reduction in the development of *macroalbuminuria*
60% reduction in the development of *clinical neuropathy*
27% reduction in the initial appearance of *retinopathy*
34-76% reduction in the *clinically significant retinopathy*
45% reduction in the progression to *severe retinopathy*
2-3 fold increase in severe hypoglycemia

type 2 diabetes. This study evaluated 110 patients with type 2 diabetes. No significant differences in hypoglycemia between the two groups was reported in this study. The results of this study are shown below in Table 2.

In addition to these studies, a great deal has been learned regarding the biochemical mechanisms of chronic complications of diabetes (e.g., glycosylation and sorbitol pathways). These advances have led to the understanding that chronic complications occur via the same mechanisms in patients with type 1 and type 2 disease and therefore that strict glycemic control is important for both populations. The results of these studies when taken together should impact not only patients with diabetes (PWD) but also the approach taken by practitioners in treating diabetes, regardless of the type of diabetes being managed.

GUIDELINES FOR GLYCEMIC CONTROL IN THE ELDERLY

The American Diabetes Association has established guidelines for glycemic control in patients with type 2 diabetes.[9] These guidelines are presented in Table 3.

In general terms the goal for most patients should be to normalize glycemic parameters. This normalization will not be possible or practical in some instances. In this type of situation, such as in the case of a very elderly patient predisposed to severe hypoglycemia, glycemic control in the acceptable range is reasonable. The above mentioned

TABLE 2

Ohkubo Study[8]–Strict Glycemic Control in Patients with Type 2 Diabetes was associated with a:

57% reduction in the development of *microalbuminuria*
100% reduction in the development of *macroalbuminuria*
69% reduction in *retinopathy*
3-4 fold higher motor and sensory nerve conduction velocity

TABLE 3. Biochemical Indices of Metabolic Control: Top Limits[9]

Biochemical Index	Normal	Goal	Action Suggested
Fasting/preprandial glucose	< 115 mg/dL (< 6.4 mM)	< 120 mg/dL (< 6.7 mM)	< 80 or > 140 mg/dL (< 4.4 mM or > 7.8 mM)
Bedtime Glucose	< 120 mg/dL (< 6.7 mM)	100-140 mg/dL (5.6-7.8 mM)	> 100 or > 160 mg/dL (< 5.6 or > 8.9 mM)
Glycosylated Hemoglobin	< 6%	< 7%	> 8%

guidelines may be adhered to with appropriate therapy which is almost always contingent of SMBG.

A long term treatment plan should be developed for all patients with diabetes and for patients > 65 years of age. The following should be considered: (1) remaining life expectancy, (2) presence of diabetic complications, (3) presence of coexisting medical or neuropsychiatric disorders, and (4) the patient's willingness and ability to comply with the proposed diabetes treatment plan.[10] One suggested method for managing elderly patients is as follows.[10] Patients with reasonable life expectancy (10-20 years), no associated medical problems or diabetes related complications, should be treated with the goal of the best possible glycemic control without undue bouts of hypoglycemia (fasting glucose 100-120 mg/dL, post-prandial glucose <180 mg/dL). Patients with lesser life expectancy, those with serious associated medical problems (cardiovascular or cerebrovascular disease), those with impaired cognition, or in those with severe microvascular disease may require more conservative therapeutic goals (fasting glucose < 140, post-prandial glucose 200-220 mg/dL).

ORAL THERAPY

The practitioner managing elderly patients with type 2 diabetes now has four different pharmacologic categories to choose from: sulfonylureas, alpha-glucosidase inhibitors, a biguanide, or a thiazolidinedione. Additionally, combinations of these oral agents may be employed as well. The following discussion will highlight efficacy and considerations for use of each of these drug categories since other topics such as mechanism of action have been covered in other manuscripts in this volume. When choice of an oral agent is being made one needs to consider the above mentioned four characteristics (life expectancy, etc.) and additionally must consider the efficacy of the medication (i.e., what level of glycemic reduction can be expected), side effects, and contraindications.

Sulfonylureas (SU)

Efficacy

SU work primarily by increasing insulin secretion.[11,12,13] Several factors that have been shown to be predictive of sulfonylurea response include age, weight, duration of disease, prior treatment with insulin, and fasting blood glucose levels.[11,12,13] Patients most likely to respond carry a recent diagnosis (disease less than 5 years), are greater than 40 years of age, weigh between 110% and 160% of ideal body weight, have fasting blood glucose of less than 200 mg/dL, and have never required insulin or needed less than 40 units of insulin per day. If these criteria are met, primary failure rates can be as low as 15%.[12] Secondary failure rates of sulfonylureas have been estimated to be 10% per year.[14] Failure rates may be much higher in some populations. A reduction in fasting plasma glucose of approximately 50-60 mg/dL and a 1-2% reduction in HbA1c can be expected in a responding patient.[15] Sulfonylureas have been shown to be effective when used in combination with acarbose, metformin, and troglitazone as will be discussed below.

Considerations in the Use of Sulfonylureas

Hypoglycemia and weight gain are the primary side effects of sulfonylureas. Overall, the incidence of hypoglycemia is variable but one study revealed a 20% chance of hypoglycemia every six months in patients treated with sulfonylureas.[14] In one major review, severe hypoglycemia was reported to be more frequent with chlorpropamide and glyburide, followed by glipizide, and finally the other first generation sulfonylureas.[14] More recently the United Kingdom Prospective Diabetes Study (UKPDS) reported the incidence of hypoglycemia after therapy with chlorpropamide, glyburide, or insulin to be 13.5%, 27.8%, and 33.4% respectively.[16] Risk factors for severe hypoglycemia include age > 60 years. Another common problem encountered with sulfonylurea therapy is weight gain. One study evaluating the effects of tolbutamide reported a mean weight increase of 1.8 kg while another study evaluating the effects of glyburide reported a mean weight increase of 2.8 (± 0.7) kg.[13,14] The UKPDS[16] reported that patients treated with chlorpropamide or glyburide experienced an average weight gain of 2.8 kg. Additionally, less common side effects include dermatologic reactions, hematologic reactions, and gastrointestinal disturbances.[14] Disulfiram-like reactions and hyponatremia have been reported with chlorpropamide.[14] Lastly, the hypoglycemic effect of sulfonylureas is mediated primarily via increases in plasma insulin concentrations. The UKPDS reported significant elevations in fasting plasma insulin concentrations in patients treated with sulfonylureas, while those treated with metformin experienced significant reductions in fasting plasma insulin concentrations. While it has been hypothesized that increases in insulin concentrations may contribute to complications, particularly cardiovascular disease, the relationship between insulin concentrations and cardiovascular disease remains to be determined.

Since all of the sulfonylureas undergo hepatic metabolism, they should be used cautiously in patients with hepatic dysfunction.[14] The active metabolite of acetohexamide, hydroxyhexamide is renally cleared. Also, 20% of chlorpropamide is excreted unchanged in the urine. Therefore, acetohexamide and chlorpropamide should not be used in patients with renal dysfunction. Also, tolazamide and glyburide have partially active metabolites that accumulate in patients with creatinine clearances of < 30 ml/min. Glipizide and tolbutamide are preferred in patients with moderate to severe renal dysfunction.[14]

In general terms, while sulfonylureas are safe to use in the elderly population, other oral agents may be more appropriate first line agents.

Lastly, a new meglitinides compound repaglinide has been approved by the FDA and is soon to be released. Its mechanisms of action is similar to sulfonylureas.

Acarbose

Efficacy

A reduction in HbA1c of 0.59%, in fasting plasma glucose of 16 mg/dL, and a 50 mg/dL reduction in postprandial plasma glucose was reported in one study evaluating the effects of acarbose monotherapy in 212 obese subjects with type 2 diabetes mellitus.[17] Similar findings were reported in another study with reported reductions in HbA1c of 0.54%, a mean reduction in fasting plasma glucose of 20 mg/dL, and a 61 mg/dL reduction in 90 minute postprandial glucose concentrations.[18]

One study evaluating the effectiveness of combination oral therapy demonstrated that acarbose was effective when used in combination with metformin, sulfonylureas, and insulin.[19] The Chaisson study reported that the addition of acarbose to insulin, metformin, or sulfonylurea therapy resulted in additional reductions in glycosylated hemoglobin levels of approximately 0.4%, 0.8%, and 0.9% respectively. These findings suggest that the effects of acarbose are additive with other agents.

It should be noted that virtually all patients will respond to acarbose. This compound has its primary effects on postprandial blood glucose levels and therefore when acarbose is employed appropriate monitoring includes the measurement of postprandial levels. Currently, acarbose is the only oral agent whose primary effect on postprandial glycemic excursions. Postprandial hyperglycemia has been shown to be toxic and should be controlled as per ADA guidelines. Acarbose is appropriate as a monotherapeutic agent in cases of mild fasting hyperglycemia. Furthermore, the addition of acarbose to any oral regimen will improve postprandial glycemic excursions and will therefore improve overall glycemic control. Another alpha glucosidase inhibitor, miglitol, has been approved by the FDA, however, it appears that this drug will not be marketed in the U.S.

Considerations in the Use of Acarbose

Acarbose is a very safe medication and is virtually free of serious side effects. The most common side effects of acarbose are gastrointestinal complaints (e.g., flatulence and diarrhea) and may in many cases be attenuated by gradual titration and continued administration of the medication. The gastrointestinal effects are dose-related. These side effects include flatulence, diarrhea, and abdominal pain which occurred in the U.S. phase III trials with an incidence greater than placebo of 45%, 21%, and 12%, respectively.[20] While elevated hepatic enzymes have been reported at higher doses (doses up to 300 mg tid), elevated serum transaminase levels were no more frequent than observed with placebo when doses of 100 mg tid or less were used.[20]

Acarbose should also be avoided in patients with inflammatory bowel disease, colonic ulceration, or obstructive bowel disorders. Because of the dose related side effects, relative contraindications to the use of this drug include chronic intestinal disorders of digestion or absorption, and patients with a medical condition that might deteriorate with increased intestinal gas formation.[20] Lastly, acarbose therapy should be avoided in patients with serum creatinine levels of > 2.0 mg/dL since studies have suggested increases in acarbose plasma concentrations with renal dysfunction and long term studies have not been carried out in this population.[20] Acarbose monotherapy does not result in hypoglycemia, however patients treated with acarbose plus any hypoglycemic agent (i.e., insulin or sulfonylureas) may experience hypoglycemia and should be counseled to self-treat any hypoglycemic episode with oral glucose.[21] Oral glucose is preferred over sucrose and other complex carbohydrates since acarbose drastically reduces the rate of absorption of complex carbohydrates.

In general terms, acarbose can be considered a first line oral agent in elderly patients with type 2 diabetes who have mild fasting hyperglycemia (glucose < 160 mg/dL) and can be used in combination with any other medication to further reduce fasting plasma glucose or to control postprandial glucose excursions.

Metformin

Efficacy

Metformin is a biguanide compound which increases glucose utilization by the muscle and reduces hepatic glucose output.[22] Metformin reduced fasting glucose levels an average of 58 mg/dl and HbA1c an average of 1.8% compared to diet-plus-placebo in one randomized, parallel, double-blinded trial of 289 moderately obese patients with type 2 diabetes mellitus.[22] Metformin is associated with other positive metabolic effects in addition to its effects on glucose. In the above mentioned trial, metformin therapy was also associated with a significant reduction in triglyceride concentrations (16%), LDL cholesterol (8%), total cholesterol (5%), and was associated with an increase in HDL cholesterol (2%). Patients in this study treated with metformin also lost a mean of 0.6 kg body weight. Metformin therapy was also associated with a significant reduction in fasting plasma insulin concentrations in the UKPDS; however the cardiovascular significance of this finding is not understood.[16]

In a second arm of the above mentioned trial, combination metformin-glyburide therapy was compared to glyburide monotherapy and metformin monotherapy in 632 obese patients with NIDDM who were previously not well controlled (FPG > 140 mg/dL) by glyburide.[22] Combination metformin-glyburide therapy reduced fasting plasma glucose by 77 mg/dL and HbA1c by 1.9% when compared to glyburide monotherapy. Patients who were not previously well controlled on glyburide monotherapy did not experience significant glycemic lowering when treated with metformin monotherapy. Metformin-glyburide therapy was also associated with a greater improvement in lipemic parameters than the metformin monotherapy or the glyburide monotherapy groups. In the combination therapy groups, reduction in triglyceride concentrations (9.2 %), LDL cholesterol (6%), total cholesterol (4.6%), and an increase in HDL cholesterol (5.4%) was observed.

A more recent double-blind, placebo-controlled, dose response trial of 451 patients with type 2 diabetes reported significant reductions in fasting plasma glucose (-78 mg/dL) and glycosylated hemoglobin concentrations (-2%) in patients treated with 2000 mg/day of metformin when compared to placebo.[23]

Another trial with metformin demonstrated that metformin monotherapy was associated with a slight weight reduction (-0.8 ± 0.5 kg) in contrast to glyburide monotherapy which was associated with a significant weight increase (+2.8 ± 0.7 kg).[24] The addition of metformin to sulfonylurea therapy attenuated the weight gain effects of the sulfonylurea and was associated with less weight gain (+0.7 ± 0.4 kg) than was observed in the glyburide monotherapy group.[24]

As mentioned above, metformin has been studied in combination with acarbose.[19] The addition of acarbose to metformin resulted in a further reduction of glycosylated hemoglobin of 0.8%. Metformin has not yet been studied in combination with troglitazone but theoretically the two agents could have synergistic effects.

Considerations in the Use of Metformin

The most common side effects of metformin are gastrointestinal. In the phase III trial in the United States, patients treated with metformin had 30% more reports of abdominal bloating, nausea, cramping, a feeling of fullness and diarrhea than patients receiving a placebo.[25] These side effects are usually self limiting, transient, and can be mitigated by starting with a low dose and titrating up slowly. These gastrointestinal effects may also be reduced by taking the medication with food. Additional less common side effects include metallic taste and a reduction in vitamin B-12 levels. Lactic acidosis can occur with the administration of metformin but is extremely rare 0.03 cases per 1000 patient years and has occurred primarily in patients with significant renal dysfunction.[25] The incidence of lactic acidosis is approximately 10 fold less than has been observed with phenformin.[26]

Metformin therapy should be closely monitored in the elderly since metformin is excreted renally and can accumulate in patients with renal dysfunction. Its use is contraindicated in patients with renal dysfunction (serum creatinine of > 1.5 mg/dL in males or > 1.4 in females) or a diminished creatinine clearance.[25] While these parameters are considered conservative, an elderly patient with little body mass could require discontinuation of the medication at lower serum creatinine levels. Lactate metabolism occurs primarily in the liver and hepatic dysfunction can lead to metabolic acidosis. Therefore metformin is contraindicated in patients with clinical or laboratory evidence of hepatic dysfunction. Metformin is also contraindicated in patients with acute or chronic lactic acidosis or a history of alcoholism or binge drinking. Lastly, metformin should also be temporarily withheld in patients with acute conditions predisposing them to acute renal failure or acidosis such as; cardiovascular collapse, acute myocardial infarction, acute exacerbation of congestive heart failure, use of iodinated contrast media, or a major surgical procedure.[25]

Metformin may be used in elderly patients with type 2 diabetes although renal function should be monitored for evidence of impairment or significant deterioration as an indication for discontinuation of therapy.[26] Because of its ability to reduce blood glucose levels, not cause weight gain, and its beneficial effect on plasma lipids and plasma insulin concentrations, metformin is a reasonable first line choice for the obese or the dyslipidemic elderly patient with moderate to severe hyperglycemia if no contraindications to the use of the drug are present.

Troglitazone

Efficacy

Troglitazone is a compound that enhances insulin utilization via its stimulation of p-par gamma receptors on the cell nucleus. Two placebo controlled trials (a 12 week and a 26 week study) evaluating troglitazone monotherapy in patients previously treated with diet reported statistically significant reductions in mean glycosylated hemoglobin concentrations (-1.4%) and in fasting serum glucose (-42 mg/dL) in patients treated with 600 mg per day when compared with placebo during the 26 week study.[27] In patients previously treated with sulfonylureas, troglitazone monotherapy did not result in an improvement in glycemic control beyond that which was observed when the patients were treated with sulfonylureas.[27]

Another trial evaluating 330 patients previously treated with diet or oral agents, randomized patients to placebo or troglitazone (doses up to 800 mg per day). At 12 weeks glycosylated hemoglobin was significantly lower in the troglitazone treated groups (7.0-7.4%) when compared to placebo (8%). Fasting plasma glucose levels were also significantly lower in the troglitazone groups (167-198 mg/dL) than in the placebo group (232 mg/dL).[28]

A trial evaluating 284 patients with type 2 diabetes who were previously poorly controlled with diet therapy randomized patients to placebo or troglitazone 400 mg/day for 12 weeks.[29] At 12 weeks glycosylated hemoglobin was significantly lower in the troglitazone treated group (8.1%) when compared to placebo (8.6%). Fasting plasma glucose levels were also significantly lower in the troglitazone group (158 mg/dL) than in the placebo group (178 mg/dL).[29]

A double-blind, placebo-controlled, 52 week study comparing combination troglitazone and micronized glyburide to micronized glyburide and to troglitazone monotherapy in 552 patients who were failing sulfonylurea monotherapy (average fasting serum glucose 224 mg/dL, glycosylated hemoglobin of > 9.6%) suggested that combination therapy was very beneficial.[27] Patients who were randomized to monotherapy with micronized glyburide or troglitazone had increases in fasting serum glucose and in glycosylated hemoglobin. Conversely patients randomized to micronized glyburide plus 600 mg per day of troglitazone had fasting serum glucose concentrations that were 79 mg/dL lower and glycosylated hemoglobin concentrations that were

2.7% lower than patients treated with glyburide monotherapy. Serum insulin concentrations were lower in the combination therapy group when compared to baseline or to the glyburide group.

Troglitazone monotherapy and combination therapy is associated with an increase in LDL (as much as 13%), HDL (up to 16%), and total cholesterol (up to 5%).[27] However, total cholesterol/HDL and LDL/HDL ratios reportedly do not change. Troglitazone does not cause an increase in ApoB fractions. Additionally, patients treated with troglitazone monotherapy or combination therapy may have reductions of up to 26% in fasting triglyceride levels as well as reductions in postprandial triglyceride concentrations.[27] Troglitazone reduces insulin resistance and its use also results in reductions in serum insulin concentrations.

Considerations in the Use of Troglitazone

Hypoglycemia has not been observed in patients treated with troglitazone monotherapy; however, its addition to other hypoglycemic medications may increase the risk of hypoglycemia.[27] In the setting of combination therapy, a reduction in the dose of the concomitant medication may be warranted when troglitazone is added to the regimen. In the North American trials of troglitazone (n = 2510 patients), 20 patients experienced liver function test abnormalities.[27] Two of these patients developed reversible jaundice with biopsies that were consistent with idiosyncratic drug reaction. One death and one liver transplant has resulted from this idiosyncratic effect. Because of this the package insert now recommends liver function tests monthly for the first six months, testing every 2 months for the next six months and periodically thereafter. Troglitazone may cause resumption of ovulation in premenopausal anovulatory patients with insulin resistance, which will be considered a positive or negative effect based on the perspective and wishes of the patient. Dose adjustment is not needed in patients with renal dysfunction. Twenty-two percent of all patients in clinical trials of troglitazone were > 65 years of age. No differences in safety or effectiveness were observed in this population when compared to younger patients. Additionally, troglitazone can cause fluid retention and has been associated with weight gain.[27]

Because of its ability to reduce blood glucose levels and plasma insulin levels, not cause weight gain, and since no dose adjustment is needed in patients with renal dysfunction, troglitazone is a reasonable first line choice in the elderly patient with moderate hyperglycemia and is a reasonable addition to other therapeutic modalities if glycemic goals have not been reached with other agents.

INSULIN

Insulin has been widely used since 1922 as monotherapy and since the late 1950s in combination therapy management of hyperglycemia in patients with type 2 diabetes mellitus. Insulin is commonly administered to elderly patients. One recent study reported that 20% of institutionalized patients with type 2 disease were treated with insulin.[30] Unfortunately, while a myriad of manuscripts have been published which delineate the metabolic effects of insulin, very few studies have directly compared the efficacy of various insulin regimens in type 2 diabetes mellitus patients with insulin monotherapy or combination therapy. Furthermore the elderly population remains virtually unstudied.

The pharmacokinetics and pharmacodynamics of conventional exogenous insulin sometimes limit its ability to attenuate the metabolic effects of type 2 diabetes, even given the vast array of insulins available today.[31] This is primarily due to the fact that no conventional preparation can provide the low insulin concentrations needed to suppress hepatic glucose production (i.e., "a basal insulin") in the postabsorptive phase and until the recent release of lispro, no preparations were able to provide the high concentrations needed to stimulate postprandial peripheral glucose disposal.

The ability of insulin therapy to lower blood glucose levels is governed by dose, regimen used, level of insulin resistance, and other factors. Insulin is effective in reducing blood sugar levels in patients with type 2 diabetes mellitus when used as monotherapy,[31,32] and when used in combination with sulfonylureas, metformin, troglitazone, and acarbose.[21]

Probably the most widely accepted combination insulin/oral therapy regimen for previously insulin-naive patients who are currently treated with oral agents is single injection intermediate acting insulin administered at bedtime. In this setting a dose of 10 U of intermediate acting insulin administered at 2200 or 2300 hours has been suggested.[33] This dose is titrated by 3-5 units q 3-4 days based on fasting blood glucose concentrations. Type 2 diabetes is a progressive disease and in time patients treated with a single daily injection may require more intensive insulin regimens.

Insulin monotherapy should be considered inadequate in most cases when the recommended ADA goals (see Table 1) are not met despite a reasonable dose of insulin.[21] What is considered a "reasonable dose of insulin" has changed over the past few years. In the recent past doses of greater than 70 units per day of insulin were considered excessive when only sulfonylureas were available and in these cases combination therapy was promoted.[33] However, given the release of oral agents which attenuate or at least mitigate insulin resistance, along with data in the past 10 years regarding insulin resistance and the possible deleterious role of hyperinsulinemia, earlier combination insulin oral therapy should be routinely considered. Perhaps all patients, including the elderly, not well controlled on more than 30 units per day should all be considered candidates for combination therapy.[21]

When combination therapy is initiated in patients currently taking insulin, the oral agent should be slowly titrated up while insulin doses are slowly titrated downward. The rapidity with which this should occur will be based on the oral agent being used and the clinical status of the patient. In many cases insulin regimens may be simplified (reduction in the number of injections) and in some cases insulin may be discontinued.

The effects of adding oral agents to insulin therapy are outlined in Table 4.

Potential Problems

Prior to the institution of insulin therapy, several possible side effects including, hypoglycemia, weight gain, and the potential for accelerated macrovascular disease should be considered in patients with type 2 diabetes mellitus. Mechanistically the elderly patient may have physical constraints which limit or even obviate their ability to use insulin products. Severe hypoglycemia, while a major concern in patients with type 2 diabetes mellitus, probably occurs with a lesser frequency than is observed in patients with type 1 disease.[32] Additionally, the counterregulatory response to hypoglycemia in patients with type 2 diabetes mellitus is blunted to a lesser degree than in patients with type 1 disease.[21] The annual rates of hypoglycemia and severe hypoglycemia (requiring the assistance of a third party) in insulin taking patients in the three year follow-up of the UK prospective study

TABLE 4. Effect of Combination Insulin/Oral Therapy

Combination	FPG reduction (mg/dL)	PPG reduction (mg/dL)	HbA1c reduction (%)	insulin dose reduction	lipid affect (+ increase, − decrease, none)
Sulfonylurea ref 34,35	−41 to −43	not reported	−0.8 to −1.1%	−24%	none
Acarbose ref 15	+1.8	−48.63 mg/dL	−0.4%	not attempted	none
Metformin ref 36	−91.5*	not reported	−1.9%	−24%	(−) TC, TG (+) HDL
Troglitazone 600 mg/day ref 34	not reported	not reported	−01.41%	−42%	(−) TG, (+) TC, LDL, HDL ξ

*avg glucose conc, ξ TC/HDL ration was unchanged

were 33.4 and 1.4%.[16] The annual rates of hypoglycemia and severe hypoglycemia in glyburide managed in this trial were 27.8 and 1.3%. One could assume that the rates of hypoglycemia would be similar or higher (if the therapy is effective) for combination insulin oral agent therapy. The incidence of any hypoglycemia in patients treated with sulfonylurea/metformin combination was reported to be 40%, combination ultralente/sulfonylurea to be 33%, and ultralente/regular insulin to be 47%.[16] However, unlike the DCCT and the UKPDS none of the cases of hypoglycemia in this study were incapacitating.

The use of insulin should not usually be considered "first line" in patients with type 2 diabetes except in cases where extreme hyperglycemia is present. Insulin may be used in combination with any of the oral agents to improve efficacy and oral agents may be added to insulin therapy with improved efficacy and often reduction in insulin dose.

Discontinuance of Insulin

Given the above mentioned adverse effects of insulin, it may be prudent to attempt stepdown therapy to oral agents in appropriate type 2 patients.

One clinical trial which compared placebo to 200 mg or 400 mg of troglitazone per day in insulin-treated patients with type 2 diabetes mellitus reported that a greater than 50% reduction in insulin doses occurred in 51% and 70% respectively in the 200 mg and 400 mg groups.[27] Insulin therapy was discontinued in 15% of patients in the 400 mg/day group and in 7% in the 200 mg/day group.

A recent study demonstrated that 77% of insulin treated patients (42 of 55 patients) with an average age of 61 yrs (±13.5) successfully discontinued insulin when treated with the combination of metformin and glyburide.[37] Hemoglobin A1c was significantly reduced (1.3% reduction) after 6 months and patients experienced continued weight loss (14.7 lbs) at 12 months. Most of the primary and secondary failures were successfully managed with combination metformin and insulin.

COST OF THERAPY

Recently an analysis of cost of therapy for patients with type 2 diabetes was published in the American Diabetes Association journal *Clinical Diabetes*.[21] As stated in this analysis, "a cursory consideration of the cost of therapy must include actual cost of the medication and cost of baseline and sequential labwork." This analysis does not include cost of self monitoring of blood glucose. Table 5 gives these parameters for the most commonly used medications. In this analysis, drug cost was based on the average wholesale price (AWP) of the drug plus an average customary pharmacy charge (15% of AWP) and on the cost of 12 monthly prescriptions. While markup varies considerably from insurance plan to insurance plan and from one pharmacy to another, this was thought to be a reasonable estimate and when applied across the board, should offer a good estimate for comparison. Insulin and syringe costs were based on patient charge applied in many chain pharmacies ($16.39/vial of Human NPH or 70/30 and $20.00/100 U-100 syringes). Laboratory costs were calculated as estimates of patient charges and of course will vary from area to area. An estimated cost of $65.00 was applied for a standard electrolyte, liver function test, lipid, and renal panel (Panel A), $80.00 for this panel plus a CBC with differential (Panel B), and $20.00 for HbA1c. The frequency of the lab tests will also of course vary widely, but in this analysis were based on ADA recommendations and strict monitoring criteria. For example, not every practitioner will feel the need for a quarterly electrolyte panel in order to assess renal function in every patient treated with metformin, howev-

TABLE 5. Cost of Therapy[21]

Drug Name and Dose	Prescription Cost/Year	Estimated Lab Cost/Year	Total
Glyburide (Brand) 10 mg/day	$501.90	Panel B X 1 = $80.00 HbA1c X 4 = $80.00	$661.90
Glyburide (Generic) 10 mg/day	$422.15	Panel B × 1 = $80.00 HbA1c × 4 = $80.00	$582.15
Glipizide (Brand) 20 mg/day	$496.62	Panel B × 1 = $80.00 HbA1c × 4 = $80.00	$656.62
Glipizide (Generic) 20 mg/day	$414.96	Panel B × 1 = $80.00 HbA1c × 4 = $80.00	$574.96
Acarbose 50 mg tid	$566.36	Panel B × 1 = $80.00 Panel A × 1 = $65.00 HbA1c × 4 = $80.00	$791.36
Troglitazone 400 mg qd	$1987.20	Panel B × 2 = $160.00 HbA1c × 4 = $80.00	$2227.20
Metformin 850 mg bid	$651.22	Panel B × 1 = $80.00 Panel A × 3 = $195.00 HbA1c × 4 = $80.00	$1006.22
Insulin NPH 10 U SQ HS (3650 units + 365 syringes)	$132.82	none in addition to labs for oral therapy	$132.82
Insulin 70/30 60 U/day in two injections (21,900 units + 730 syringes)	$504.00	none in addition to labs for oral therapy	$504.00

er, in many cases and particularly in the case of an elderly patient, this would be a prudent course of action. The reader is encouraged to evaluate the cost table in the context of their own practice. Cost of each of these entities was expressed in dollars per year of therapy.

It is very important that when drug cost is considered, several other factors are examined as well. The total cost of diabetes is estimated to exceed $90 billion dollars per year and it accounts for one in every seven dollars spent on health care in the nation, even though only 6% of the U.S. population has diabetes.[38] Of this $90 billion dollars per year, 42.5% is used for treating inpatients in hospitals and nursing homes. Only 6.7% of the cost of treating diabetes is directed towards outpatient treatment. Medication cost accounts for only 1.3% of the total cost of treating diabetes. Unfortunately, our medical care system remains a reactive entity, treating major inpatient problems at a great cost, rather than functioning in a more proactive fashion, expending reasonable amounts of resources up front, obviating many inpatient problems, and reducing mortality rates. While cost of medications needs to be considered, the most salient question is "what is the best drug for a given patient?" Given our current armamentarium of antidiabetic medications, their different metabolic effects, side effect profiles, efficacy, and contraindications, along with overwhelming knowledge that strict glycemic control translates into significantly lower morbidity and mortality, the cost of medication becomes almost negligible. If patients are treated in a proactive preventative fashion with the best medication for their metabolic and physical profile, morbidity and mortality will be reduced, patients will have greater quality of life, and the health care system will save enormous amounts of money, regardless of the medication used.

CHOICE OF TREATMENT REGIMEN

In very general terms type 2 diabetes can be treated in a stepwise fashion starting with step I–diet and exercise, moving next to step II–oral monotherapy (SU, metformin, acarbose, troglitazone, or repaglinide), next step III–oral combination therapy (SU, metformin, acarbose, troglitazone, or repaglinide), step IV–combination oral agent/insulin therapy, next step V–insulin monotherapy, and lastly, if insulin monotherapy exceeds 30 units per day step VI–the addition of an oral agent reducing insulin resistance may be appropriate. Also, step down therapy may be considered to get the patient back on oral therapy, particularly if equivalent or better glycemic control can be achieved. A recently published algorithm suggested that several factors be considered when choosing an oral agent or an oral agent combination, including efficacy of the medication (blood glucose lowering potential), current and goal blood glucose and glycosylated hemoglobin levels, contraindications, side effects, and effect on other metabolic parameters such as insulin concentration, lipids, and body weight.[39] Table 6 summarizes key considerations in the use of oral therapeutic agents.

TABLE 6. Effects of Oral Monotherapy and Oral Combination Therapy

Regimen	FPG reduction (mg/dL)	PPG reduction (mg/dL)	HbA1c reduction (%)	serum insulin concentration	lipid affect (+ increase, − decrease, none)	body weight
sulfonylureas (SU) (ref 39)	50-60	not reported	1-2% reduction	significant increases	no direct effect	increase
SU and acarbose (ref 19)	32 mg/dL lower than in the SU monotherapy group	73 mg/dL lower than in the SU monotherapy group	0.9% lower than in the SU monotherapy group	no sig effect	no sig effect	not reported
SU and metformin (ref 22)	77 mg/dL lower than in the SU monotherapy group	61 mg/dL reduction compared to SU monotherapy *	1.9% reduction compared to SU monotherapy group	no sig change	−TC, LDL, TG, + HDL compared to SU monotherapy group	0.4 kg increase
SU and troglitazone (600 mg/day) (ref 27)	79 mg/dL lower than in the SU monotherapy group	not reported	2.7% reduction compared to SU monotherapy group	significant reduction	+TC, LDL, HDL, −TG**	no change
acarbose monotherapy (ref 18)	20 mg/dL	61 mg/dL	0.54% reduction	significant reduction	−TG, LDL, TC, +HDL	not reported
acarbose and metformin (ref 19)	23 mg/dL less than in the metformin monotherapy group	62 mg/dL reduction when compared to metformin monotherapy	0.8% lower than in the metformin monotherapy group	no sig effect	no sig effect	not reported
metformin monotherapy (ref 22,23)	78 mg/dL	72 mg/dL reduction	2%	no change	−TC, LDL, TG, +HDL	0.6 kg wt loss
troglitazone monotherapy (600 mg/day) (ref 27)	42 mg/dL** reduction compared to placebo	not reported	1.4%	sig reduction	+TC, LDL, HDL, −TG**	no change

* reduction in mean plasma glucose concentration after oral glucose, ** TC/HDL ratio was unchanged

CONCLUSION

The incidence of type 2 diabetes mellitus among the elderly is escalating. The options available for practitioners treating these patients has grown exponentially in the past few years. While none of the currently available medications used in the management of diabetes is contraindicated in the elderly, some regimens are preferred over others. Practitioners should establish individualized glycemic goals for each patient based on there age, co-morbid conditions, and other factors. Once goals are established the practitioner should then proceed in a stepwise fashion to reach these goals. The cost of medication may be considered in the choice of a particular regimen, however cost of medication is incidental when compared to the savings realized with appropriate glycemic and metabolic control.

REFERENCES

1. Diabetes Across the Lifespan 2nd edition, Joshu-Haire D ed. Diabetes Mellitus and the Older Adult. Funnell MM, Merritt JH. Mosby, St. Louis. 1996.

2. Halter J. chapter. Geriatric Patients in Lebvitz H ed. Therapy for Diabetes Mellitus and Related Disorders, 2nd edition. American Diabetes Association. Alexandria. 1994.

3. Wilson PW, Anderson KW, Kannel WB: Epidemiology of Diabetes in the Elderly: The Framingham Study, American Journal of Medicine. 1986;80(5A):3-9.

4. Miller LV: Managing the Elderly Patient with Diabetes. Clinical Diabetes 1985;3(4):73-78.

5. U.S. National Commission on Diabetes: Report of the National Committee on Diabetes to the Congress of the U.S. Dept. of Health, Education & Welfare. Public Health Service, NIH, 1976, 4 Volumes. (DHEW Pub. (#NIH) 76-1018, 76-1022, 76-1031,76-1033).

6. The Diabetes Control and Complications Trial Research Group. The Effect of Intensive Treatment of Diabetes on the Development and Progression of Long-term Complications in Insulin-Dependent Diabetes Mellitus. N Engl J Med. 1993; 329;14:977-986.

7. Klein R. Hyperglycemia and microvascular and macrovascular disease in diabetes. Diabetes Care. 1995;18:258-268.

8. Ohkubo Y, Kishikawa H, Araki E, Miyata T, Isami S, et al. Intensive insulin therapy prevents the progression of diabetic microvascular complications in Japanese patients with non-insulin-dependent diabetes mellitus: a randomized prospective 6-year study. Diabetes Res Clin Prac. 1995;28(2):103.

9. Standards of Medical Care for Patients with Diabetes Mellitus. Diabetes Care. 1998;21;supp 1:S23-S31.

10. Geriatric Patients, Halter J. Chapter in: Therapy for Diabetes Mellitus and Related Disorders. Second Edition, American Diabetes Association. Alexandria, 1994.

11. Leahy JL, and others: Chronic hyperglycemia is associated with impaired glucose influence on insulin secretion: a study of normal rats using chronic in vivo glucose infusions, J Clin Invest, 77:908-915,1986.

12. Lebovitz H: Clinical utility of oral hypoglycemic agents in the management of patients with non-insulin-dependent diabetes mellitus, Am J Med 75 (suppl 5B): 94-99, 1983.

13. Jackson JE and Bressler R: Clinical pharmacology of sulfonylurea hypoglycaemic agents, Drugs 22:211-245; 295-320,1981.

14. Gerich J.E. Oral Hypoglycemic Agents. NEJM 321:1232-1245, 1989.

15. Lebovitz H. A New Oral Therapy for Diabetes Management: Alpha-Glucosidase Inhibition with Acarbose. Clinical Diabetes. 13:99-102,1995.

16. United Kingdom Prospective Diabetes Study Group. United Kingdom Prospective Diabetes Study (UKPDS) 13: relative efficacy of randomly allocated diet, sulphonylurea, insulin, or metformin in patients with newly diagnosed non-insulin dependent diabetes followed for three years. BMJ. 1995;310:83-8.

17. Conniff RE, Shapiro JA, Seaton TB. Long-term Efficacy and Safety of Acarbose in the Treatment of Obese Subjects with Non-Insulin-Dependent Diabetes Mellitus. Arch Intern Med. 1994; 154:2442-2448.

18. Coniff RF, Shapior JA, Seaton TB, Bray G. Multicenter, Placebo-Controlled Trial Comparing Acarbose (BAY g5421) with Placebo, Tolbutamide, and Tolbutamide-Plus-Acarbose in Non-Insulin-Dependent Diabetes Mellitus. American Journal of Medicine 98:443-451,1995.

19. Chiasson J-L, Josse RG, Hunt JA, Palmason C, et al. The Efficacy of Acarbose in the Treatment of Patients with Non-Insulin Dependent Diabetes Mellitus. Annals of Internal Medicine, 1994;121;12:928-935.

20. Precose™ package insert, Bayer Pharmaceuticals, West Haven, CT.

21. White J. Combination Oral Agent/Insulin Therapy in Patients with Type II Diabetes Mellitus. Clinical Diabetes. 1997;15;3:102-113.

22. DeFronzo RA, Goodman AM, and the Multicenter Metformin Study Group. Efficacy of Metformin in Patients with NIDDM. N Engl J Med. 1995;333:541-549.

23. Garber AJ, Duncan TG, Goodman AM, Mills DJ. Efficacy of Metformin in Type II Diabetes: Results of a Double-Blind, Placebo-controlled, Dose-response Trial. Am J Med (in press).

24. Hermann LS, Schersten B, Bitzen P, Kjellstrom T, Lindgarde F, Melander A. Therapeutic Comparison of Metformin and Sulfonylurea, Alone and in Various Combinations. Diabetes Care 17;10:1100-1109,1994.

25. Glucophage package insert, Bristol-Meyers Squibb.

26. Blonde L, Guthrie RD, Sandberg MI. Metformin: An Effective and Safe Agent for Initial Monotherapy in Patients with Non-Insulin-Dependent Diabetes Mellitus. The Endocrinologist. 1996;6;6:431-438.

27. Rezulin package insert. Parke-Davis, Morris Plains, New Jersey.

28. Kumar S, Boulton AJM, Beck-Nielsen H, Berthezene F, et al. Troglitazone, an insulin action enhancer, improves metabolic control in NIDDM patients. Diabetalogia. 1996;39:701-709.

29. Iwamoto Y, Kosaka K, Kuzuya T, Akanuma Y, Shigeta Y, Kaneko T. Effects of Troglitazone. Diabetes Care. 1996;19;2:151-156.

30. Benbow SJ, Walsh A, Gill GV. Diabetes in institutionalized elderly people: a forgotten population? BMJ 1997;314:1868-1869.

31. Galloway J.A. Treatment of NIDDM with Insulin Agonists or Substitutes. Diabetes Care. 13;12:1209-1239. 1990.

32. Genuth S. Insulin Use in NIDDM. Diabetes Care. 13;12:1240-1264. 1990.

33. Therapy for Diabetes Mellitus and Related Disorders, 2nd ed. Lebovitz H ed. Combination Therapy for Hyperglycemia, Lebovitz H. American Diabetes Association, Alexandria, VA. 1994.

34. Pugh JA, Wagner ML, Sawyer J, Ramirez G, Tuley M, Friedberg SJ. Is Combination Sulfonylurea and Insulin Therapy Useful in NIDDM Patients? A metaanalysis. Diabetes Care. 1992;15;8:953-959.

35. Johnson JL, Wolf SL, Kabadi UM. Efficacy of Insulin and Sulfonylurea Combination Therapy in Type II Diabetes, A meta-analysis of randomized, placebo-controlled studies. Arch Int Med. 1996; 156:259-264.

36. Gugliano D, Quatraro A, Consoli G, Minei A, Ceriello A, DeRosa N, Onofrio ED. Metformin for obese, insulin-treated diabetic patients: improvement in glycaemic control and reduction of metabolic risk factors. European Journal of Clinical Pharmacology. 1993;44:107-112.

37. Bell D, Mayo M. Outcome of Metformin Facilitated Reinitiation of Oral Diabetic Therapy in the Insulin Requiring Non-Insulin-Dependent Diabetic Patient. (to be published in April in Endocrine Practice).

38. American Diabetes Association, ed. Diabetes 1996 Vital Statistics. 1996, American Diabetes Association. Rockville, Maryland.

39. White J. The Pharmacologic Management of Patients with Type II Diabetes Mellitus in the Era of New Oral Agents and Insulin Analogs. Diabetes Spectrum. 1996;9;4:227-234.

Acarbose: 'The European Experience'

Maria M. Byrne
Moria A. Burhorn
Burkhard Göke

SUMMARY. Acarbose, an α-glucosidase inhibitor, delays carbohydrate digestion, and thereby decreases postprandial blood glucose levels. It is a useful drug especially for the treatment of obese subjects with type 2 who fail first-line therapy with diet and exercise. It can be used as first line therapy in diabetic individuals in whom postprandial hyperglycemia is significantly greater than fasting hyperglycemia, and in elderly patients with mild diabetes who are at high risk for hypoglycemia. It causes reductions in HbA1c of 0.5-1% and reductions of postprandial blood glucose levels of 50 mg/dL. No weight gain has been described with acarbose and it is effective with high carbohydrate intake. It is also useful in combination with other oral hypoglycemics and insulin. In type 1 subjects, acarbose may reduce glycemic fluctuations and reduce insulin dosage. Acarbose may reduce episodes of mid-evening and nocturnal hypoglycemia. The disadvantage of acarbose is its gastrointestinal side-effects of flatulence and diarrhoea which can be minimized by slowly titrating the doses according to the patient's symptoms. *[Article copies available for a fee from The Haworth Document Delivery Service: 1-800-342-9678. E-mail address: getinfo@haworthpressinc.com]*

KEYWORDS: diabetes, acarbose, elderly

Maria M. Byrne, Moria A. Burhorn, and Burkhard Göke are affiliated with the Department of Internal Medicine, Clinical Research Unit for Gastrointestinal Endocrinology, Philipps University Marburg, Baldingerstr., 35033 Marburg, Germany.

[Haworth co-indexing entry note]: "Acarbose: 'The European Experience.'" Byrne, Maria M., Moria A. Burhorn, and Burkhard Göke. Co-published simultaneously in *Journal of Geriatric Drug Therapy* (Pharmaceutical Products Press, an imprint of The Haworth Press, Inc.) Vol. 12, No. 2, 1999, pp. 47-59; and: *Diabetes Mellitus in the Elderly* (ed: James W. Cooper) Pharmaceutical Products Press, an imprint of The Haworth Press, Inc., 1999, pp. 47-59. Single or multiple copies of this article are available for a fee from The Haworth Document Delivery Service [1-800-342-9678, 9:00 a.m. - 5:00 p.m. (EST). E-mail address: getinfo@haworthpressinc.com].

© 1999 by The Haworth Press, Inc. All rights reserved.

INTRODUCTION

Non-insulin-dependent diabetes mellitus (type 2) is a common disease associated with high morbidity and mortality from macrovascular and microvascular complications.[1-4] The current high incidence of complications in patients with type 2 indicates that the current therapeutic approach is inadequate for maintaining good health.[3,5] Type 2 occurs due to development of a combination of insulin resistance and β-cell failure. The major therapeutic goals in patients with type 2 are to normalize blood glucose levels, reduce obesity, and normalize lipid disturbances and increased blood pressure, in order to improve the well-being of the patient and to reduce the risk of developing late diabetic complications. Results of the Diabetes Control and Complications Trial (DCCT) demonstrate that intensive glycemic control in type 1 patients will prevent the progression of at least the microvascular complications like retinopathy and nephropathy.[6] To date there are no long term studies in type 2 patients to show that treatment with oral antihyperglycaemic agents helps to postpone or prevent complications. It is expected that we can extrapolate from the results of the DCCT[7] and assume that normalizing glucose concentrations also makes sense in subjects with type 2 diabetes.

Acarbose, an α-glucosidase inhibitor is a drug for the treatment of adult diabetic patients in combination with diet and with other oral anti-diabetic drugs or insulin.[8] Acarbose delays carbohydrate digestion by competitively inhibiting α-glucosidases such as glucoamylase, sucrase, maltase, and isomaltase in the small intestine. However acarbose does not affect total glucose absorption. By this mechanism acarbose is able to flatten blood glucose concentrations following carbohydrate ingestion. The therapeutic potential of acarbose in the treatment of both type 2 and type 1 diabetes has been extensively studied in Europe and elsewhere. Here we will discuss predominantly the European experience.

USE OF ACARBOSE IN DIABETICS FAILING DIETARY THERAPY

Acarbose in addition to diet treatment has been shown to be more effective than diet alone.[9,10,11] In 1991 Hanefeld et al. treated subjects with insufficiently diet treated type 2 diabetes (mean HbA1c of 9.2%)

with acarbose 100 mg tid for 6 months.[12] Both fasting, and 1-hour postprandial blood glucose levels were lowered by 1 mmol (18 mg/dL) and 3 mmol (54 mg/dL), respectively, in the acarbose treated group compared to placebo. Subjects had an average reduction of 1% in the HbA1c levels (9.3% to 8.65%). This is consistent with the results from other non-European multicenter studies.[13] In addition to the glucose lowering effects 5-h postprandial insulin levels were reduced in the acarbose treated group.

There was a significant reduction in the 1-hr postprandial triglyceride levels, with no change in the total cholesterol values. Body weight was completely unchanged with acarbose therapy, in this study. When nutrient intake was studied in subjects treated with acarbose there was no change in the intake of energy and carbohydrates, but there was a slight decrease in dietary fat intake but no change in total body weight.[14]

These results suggest that acarbose could be used as a first-line agent for the obese, hyperinsulinemic patient with mildly elevated HbA1c levels or 2-hr postprandial glucose levels. The question arises as to whether acarbose should be used instead of/or in combination with other established oral drugs. Acarbose 100 mg tid has been compared to glibenclamide 4.3 mg/day as first-line agents for insufficiently diet treated subjects with type 2 disease (HbA1c levels 7-9%). Subjects were treated for 6 months, and both agents decreased fasting blood glucose levels, 1-hr postprandial, and HbA1c levels by similar magnitude compared to placebo. The mean relative insulin increase 1-hr postprandially was 1.5 in the placebo group, 1.1 in the acarbose group, and 2.5 in the glibenclamide group.[15] So acarbose and glibenclamide are equally effective as monotherapy of type 2 disease when diet alone fails. As hyperinsulinemia is associated with increased risk for cardiovascular disease, acarbose has the added benefit of not elevating postprandial insulin levels. Both drugs caused no weight gain, over six months.[15] Acarbose has been shown to be similar, or slightly less effective, when compared to sulphonylureas[16,17] or metformin[18] as monotherapy.

USE OF ACARBOSE IN SUBJECTS FAILING TREATMENT WITH OTHER ORAL HYPOGLYCEMIC AGENTS

Acarbose has been shown to be effective in subjects poorly controlled by diet and sulphonylureas.[19,20] In fact the addition of acarbose may be useful to gain some time before beginning insulin therapy. Acarbose in addition to sulphonylureas led to significant reduction in postprandial blood glucose concentrations (approximately 3 mmol; 54 mg/dL), and decreases in glycosylated hemoglobin levels of 0.4-0.6%.[21,22] The plasma insulin response to meals is not improved by acarbose therapy. In a recent Spanish study it was shown that in subjects with secondary sulphonylurea failure both insulin and metformin plus sulphonylureas provided better glycemic control than acarbose plus sulphonylureas. Metformin combined with sulphonylureas offered further advantages for the control of body weight and blood pressure.[23]

In a North American study, Reaven et al.[24] made an interesting observation that there was considerable individual variation in response to acarbose in subjects failing sulphonylurea therapy. Some 4 of 12 subjects had a dramatic response with reduction in both fasting and 2-h postprandial glucose levels, and 4 of 12 subjects had essentially no response, suggesting that the addition of acarbose may be useful in a selective group of subjects.

USE OF ACARBOSE IN SUBJECTS WITH TYPE 2 DISEASE ON INSULIN THERAPY

Acarbose was shown to have a small effect in insulin treated type 2 patients with a drop in HbA1c of 1.5% after 6 months compared to a drop of 0.9% in the placebo group.[25] An insulin-acarbose combination may be useful in subjects with persistent elevation in postprandial glucose levels on insulin alone[26] and in fact may reduce the insulin requirements for the control of postprandial hyperglycemia. These results are consistent with those found in the North American study, with a 24-week double-blind treatment period, with forced titration of acarbose from 50 mg tid to 300 mg tid, in conjunction with diet and insulin therapy. HbA1c levels were reduced by 0.4% (P = 0.0001) and the total daily insulin dose by 8.3% (P = 0.0015), fasting glucose levels were reduced by 0.9 mM (16.2 mg/dL) (P = 0.044) with a 2.6 mM (46.8 mg/dL) reduction in peak postprandial glucose levels (P = 0.0001). These results suggest that acarbose is a safe and effective adjunct to diet and insulin therapy for the management of insulin-requiring type 2 subjects.[27] The additive acute effect of a single dose of acarbose, 100 mg at breakfast in 36 insulin dependent type 2 subjects and 10 type 1 subjects, was demonstrated after 7 days of treatment by a marked flattening of elevated postprandial morning blood glucose profiles regardless of the type of diabetes.[28]

Recently it has been reported that after two years of treatment with acarbose, the reduction in HbA1c levels (1.2%) persisted in both type 1 and type 2 subjects.[29] Longer term studies are required to establish if this effect further persists. The acarbose arm of the United Kingdom Prospective Diabetes Study will study the effect of acarbose over a longer period of time.[30] Results after one year of this trial demonstrated that acarbose caused a 0.6% reduction in HbA1c levels, with no increased no. of hypoglycemic episodes and no change in concurrent medications. The compliance in this study was only 50% after one year as patients discontinued the acarbose secondary to flatulence and diarrhoea. The compliance in the placebo group was 72%.[31]

USE OF ACARBOSE IN SUBJECTS WITH TYPE 1 DIABETES MELLITUS

Acarbose causes a reduction in postprandial blood glucose levels and hence a smoothing of the daily blood glucose profiles, leading to a reduction in HbA1c levels and a decrease in insulin requirements. Raptis et al. was first to study the effect of acarbose in subjects with type 1 disease and to use the artificial pancreas to measure blood glucose levels for 24 hours. They demonstrated a clear reduction in postprandial glucose levels and an associated reduction in insulin requirements.[32,33] These results were confirmed in other short-term studies.[19,34] Dimitriades et al. demonstrated that the glucose lowering effects of acarbose occurred independent of the timing of the insulin injection.[35] One of the most carefully conducted studies of the effect of acarbose in type 1 subjects was performed by Gerard et al.[36] This was a single blind study which lasted 6 months in 32 ambulatory type 1 subjects. The study consisted of three periods of two months each. During the first and the third period the subjects used twice daily insulin and placebo three times daily. During the second period the subjects used twice daily insulin and acarbose 100 mg tid. Neither caloric intake nor body weight changed during the three study periods. The mean amplitude of postprandial glycemic excursions decreased during acarbose therapy and not during placebo. A small but statistically significant decrease in insulin requirements was observed during acarbose therapy. HbA1c levels and postprandial C-peptide responses remained unchanged. Moderate hypoglycemia was reported more often during acarbose therapy.

Other studies have confirmed that the addition of acarbose to insulin in subjects with type 1 causes some reduction in insulin requirements and a small reduction in HbA1c levels.[37-39] This reduction in postprandial glucose values has been associated with an increase in hypoglycemic episodes. In contrast, a study by McCullogh et al. has shown that 100 mg of acarbose with the evening meal may prevent mid-evening and nocturnal hypoglycaemia reactions in type 1 subjects. In the same study 200 mg of acarbose exerted a more profound effect and caused more mid-evening hypoglycemia.[40] Most of the multicentre studies performed demonstrate an improvement in metabolic control in patients with type 1 diabetes.[40,41,42]

EFFECT OF ACARBOSE ON LIPIDS IN TYPE 2 DIABETES MELLITUS

In a double-blind, cross-over, placebo-controlled randomized study in poorly controlled type 2 subjects treated with sulphonylureas, acarbose caused significant reductions in postprandial glucose levels and, in addition, a decline in triglyceride levels which was not statistically significant.[43] In a randomized double-blind placebo-controlled study of 94 subjects with type 2, treated, after a run in period of 3 months of diet alone, with 100 mg of acarbose tid for 24 weeks, no significant decrease in fasting total cholesterol, HDL cholesterol, or triglycerides between the acarbose treated group and placebo was found. Acarbose treatment resulted in a significant reduction in postprandial hypertriglyceridemia ($P < 0.01$) 1-h after a test meal of 400 kcal (50% carbohydrate, 35% fat, 15% protein). In this study acarbose reduced postprandial blood glucose levels and the insulin increment, but this was not associated with a change in free fatty acid levels. It is possible that the correction of hyperinsulinemia may contribute to a reduction in VLDL-triglyceride production from the liver.[44]

In the subset of subjects with initial total cholesterol values of >260 mg/dL the mean decrease in total cholesterol during acarbose therapy was 14%. In this group the ratio of total cholesterol to HDL cholesterol decreased during acarbose ($P < 0.0025$). LDL cholesterol increases slightly in both acarbose and placebo groups. Subjects with initial total cholesterol levels <260 mg/dL, showed no significant reduction in total cholesterol levels. Other short-term or non-controlled trials have shown a small reduction in serum triglycerides with no effect on total cholesterol levels.[24,45] In general, acarbose does not seem to have primary effects on either triglyceride or cholesterol metabolism.

USE OF ACARBOSE IN THE TREATMENT OF REACTIVE HYPOGLYCEMIA

The first therapeutic approach of reactive hypoglycemia is dietary, i.e., small regular feeding low in carbohydrates, avoidance of drinks rich in sucrose or glucose or drinks containing sugar and alcohol, and addition of soluble fibre to the diet. When symptoms persist they may be improved by the use of acarbose.

In a double-blind cross-over study of acarbose in subjects with functional hypoglycemia, Gerard et al. demonstrated that acarbose reduced the magnitude of post-sucrose reactive hypoglycaemia and reduced the plasma insulin rise in response to sucrose ingestion.[46] These results were confirmed by Richard et al., where acarbose reduced the blood glucose and plasma insulin rises in response to an oral sucrose load. In fact the blood glucose values between 150 min to 240 min were higher after acarbose.[47]

Acarbose has also been shown to be useful in the treatment of the dumping syndrome. Subjects reported symptomatic improvement, when acarbose was given with a 50 g sucrose meal. Acarbose resulted in a reduction of the hyperglycemic phase and hence improvement of the hypoglycemic phase.[48] Long-term studies are necessary to establish this indication for acarbose therapy.

EFFECT OF ACARBOSE ON GLP-1 SECRETION

In normal subjects, acarbose has been shown to delay the absorption of orally administered sucrose leading to prolonged GLP-1 responses at maximal postprandial levels.[49] GLP-1 is an insulinotrophic peptide hormone that is released from the L-cells of the ileum and colon in response to the ingestion of carbohydrate, protein or fat. The physiological effect of the elevated GLP-1 plasma levels is not known but is being actively investigated.

EFFECT OF ACARBOSE ON COAGULATION

Diabetes mellitus is a thrombosis-prone state and hyperglycemia activates hemostasis. Postprandial hyperglycemia raises 2 markers of coagulation activation, prothrombin fragments 1 and 2 and D-dimers.[50,51] Seventeen diet treated type 2 subjects received 100 mg acarbose or placebo before a standard meal, and acarbose was shown to significantly reduce the rise in glucose, insulin, prothrombin fragments 1 and 2 and D-dimers from 0-240 minutes post ingestion. By decreasing post meal hyperglycaemia, acarbose may decrease meal-induced activation of hemostasis in diabetic patients.[52] The ultimate benefit for preventing thromboembolic complications is speculative.

TOLERABILITY OF ACARBOSE

The tolerability of acarbose was examined in a postmarketing surveillance survey of 10,462 patients (829 with type 1, 9,440 with type 2, 193 not classified) during a 12 week period. Of the patients, 78.6% had no adverse events, 19% reported meteorism/flatulence and 3.2% diarrhoea. Hypoglycemia was found in 0.8% of type 1 and 0.6% of type 2 who were receiving concurrent insulin (n = 8) or glibenclamide (n = 1) treatment. There was no indication of other adverse effects, i.e., elevated levels of transaminases or creatinine.[53] When the acarbose dose is tailored to individual needs and tolerability, the adverse effects are reduced. The currently recommended acarbose dosing schedule is now 25 mg per day with the first bite of the largest meal for one week, then 25 mg tid for one week, then 25 mg tid for 2 weeks gradually titrating up to 50 mg tid by week 7 and 100 mg tid by week 12[54] dependent on individual needs and tolerability. Liver transaminases and creatinine must be monitored during the course of therapy. Patients should be informed about the side effects prior to starting the drug. Previous American studies have reported more adverse events and reversible elevations in liver transaminases but the dosages used were higher.[13] To clarify the optimum timing for ingestion of acarbose, a 100 mg dose was administered 30 min before, at the beginning, and 15 min after ingestion of a test meal. The smallest increases in blood glucose occurred when acarbose was taken with the first mouthful of food.[55]

INTERACTION BETWEEN ACARBOSE AND OTHER HYPOGLYCEMIC AGENTS

It has been shown that acarbose does not significantly modify the pharmacokinetic characteristics of glibenclamide, one of the most frequently prescribed sulphonylurea compounds,[56] so acarbose can be safely administered with sulphonylureas. In contrast, it has been shown that acarbose significantly reduces the acute bioavailability of metformin in normal subjects. The effect is predominant during the first 3-h following the combined ingestion of both compounds, but the overall 24-h bioavailability is not significantly altered.[57] A recent study has shown that in patients with type 2 diabetes and unsatisfactory acarbose monotherapy good metabolic control can safely be achieved by combination with metformin.[58] Clinical studies of combination therapy with troglitazone and acarbose are in progress.

CONCLUSION

Acarbose is a useful drug especially for the treatment of obese subjects with type 2 diabetes who fail first-line therapy with diet and exercise. It may also be useful when added to the other available medications, sulphonylureas, metformin, troglitazone or insulin. Long term prospective studies are required to establish if it should be used as first-line therapy. To date the advantages of acarbose over sulphonylureas or insulin are that it does not aggravate hyperinsulinemia commonly present in subjects with type 2 diabetes. This may be advantageous in preventing the development of atherogenesis in the long term.[59] Acarbose also causes some lowering of plasma triglyceride levels. No weight gain has been described with acarbose therapy thereby causing no worsening of insulin resistance. When used as first-line therapy, acarbose does not cause hypoglycaemia and therefore may be useful in elderly patients with mild diabetes and other subjects who are at high risk for hypoglycaemia. The other advantage of acarbose is that it is still effective during high carbohydrate intake.

In subjects with type 1 diabetes, addition of acarbose can reduce glycaemic fluctuations, and reduce insulin dosage. Acarbose can reduce the number of hypoglycaemic episodes if the insulin dose is appropriately adjusted or if it is taken before the evening meal it can reduce the episodes of mid-evening and nocturnal hypoglycaemia.

The disadvantage of acarbose is its gastrointestinal side-effects of flatulence, and diarrhoea which can be minimized by starting with a low dose and slowly titrating the dose according to the patients symptoms.

REFERENCES

1. Garcia MJ, McNamara PM, Gordon T, Kannell WB. Morbidity and mortality in diabetics in the Framingham population. Sixteen year follow-up study. Diabetes. 23:105-111, 1974.

2. Stamler J, Vaccaro O, Neaton JD, Wentworth D. Diabetes, other risk factors, and 12-yr cardiovascular mortality for men screened in the Multiple Risk Factor Intervention Trial. Diabetes Care 16:434-444, 1993.

3. Panzram G. Mortality and survival in type 2 (non-insulin-dependent) diabetes mellitus. Diabetologia 30:123-131, 1987.

4. Walters DP, Gatling W, Houston AC, Mullee MA, Julious SA, Hill RD. Mortality in diabetic subjects: an eleven-year follow-up of a community-based population. Diabet Med 11:968-973, 1994.

5. Knatterud GL, Klimt CR, Levin ME, Jacobson ME, Goldner MG. Effects of hypoglycaemic agents on vascular complications in patients with adult onset diabetes VII. Mortality and selected nonfatal events with insulin treatment. JAMA 240:37-42, 1978.

6. The Diabetes Control and Complications Trial Research Group. The effect of intensive treatment of diabetes on the development and progression of long-term complications in insulin-dependent diabetes mellitus. N Eng J Med 329:977-986, 1993.

7. Nathan DM. Inferences and implications. Do results from the Diabetes Control and Complications Trial apply in NIDDM? Diabetes Care 18:251-257, 1995.

8. Puls W, Keup U, Krause HP et al. Pharmacology of an α-glucosidase inhibitor. Front Hormone Res 7:235-247, 1980.

9. Schumann F. Langzeitbehandlung mit Acarbose. Med Welt 33:1686, 1981.

10. Rosenkranz R, Hillebrand I, Böhme K. Improvement of carbohydrate metabolism by acarbose in maturity onset diabetes, not sufficiently treated with diet only. In: Creutzfeldt W ed., First International symposium on acarbose, 305, 1982.

11. Folsch UR. Clinical experience with acarbose as first line therapy in NIDDM. Clin. Invest. Med. Aug, 18(4): 312-317, 1995.

12. Hanefeld M, Fischer S, Schulze J, Spengler M, Wargenau M, Schollberg K, Fücker K. Therapeutic potentials of acarbose as first-line drug in NIDDM insufficiently treated with diet alone. Diabetes Care 14:732-737, 1991.

13. Coniff RF, Shapiro JA, Seaton TB. Long-term efficacy and safety of acarbose in the treatment of obese subjects with non-insulin-dependent diabetes mellitus. Arch Intern Med 154(12) 2442-2448, 1994.

14. Tuomilehto J, Pohjola M, Lindstrom J, Aro A. Acarbose and nutrient intake in non-insulin dependent diabetes mellitus. Diabetes Res Clin Pract 26(3):215-222, 1994.

15. Hoffmann J, Spengler M. Efficacy of 24-week monotherapy with acarbose, glibenclamide, or placebo in type 2 patients. Diabetes Care 17;561-566, 1994.

16. Fölsch UR. Efficacy of glucosidase inhibitors compared to sulphonylureas in the treatment and metabolic control of dietary treated type 2 diabetics. Diabetes Nutr Metab 1990;3 (Suppl 1):63-68, 1990.

17. Mies R, Spengler M. Efficacy of the glucosidase inhibitor acarbose compared to the sulphonylurea glisoxepid on metabolic control of type 2 (non-insulin-dependent) diabetes (Abstract). Diabetologia 30:557A 1987.

18. Schwedes U, Petzoldt R, Hillebrand I, Schöffling K. Comparison of acarbose and metformin treatment in non-insulin-dependent diabetic outpatients. In: Creutzfeldt W, ed. Proceedings of the First International Symposium on Acarbose. Amsterdam: Excerpta Medica, 275-281, 1982.

19. Sachse G, Willms B. Effect of the α-glucosidase inhibitor BAY g 5421 on blood glucose control of sulphonylurea-treated diabetics and insulin treated diabetics. Diabetologia 17:287-290, 1979.

20. Willms B. Acarbose in non-insulin-dependent diabetes: Short-term studies in combination with oral agents. In: Creutzfeldt W, ed. Acarbose for the treatment of Diabetes Mellitus. Berlin: Springer, 79-91, 1988.

21. Clissold SP, Edwards C. Acarbose: a preliminary review of its pharmacodynamic and pharmacokinetic properties, and therapeutic potential. Drugs 35:214-243, 1988.

22. Sachse G. Acarbose in non-insulin-dependent diabetes. Long-term studies in combination with oral agents. In: Creutzfeldt W, ed. Acarbose for the treatment of Diabetes Mellitus. New York: Springer, 92-111, 1988.

23. Calle-Pascual AL, Garcia-Honduvilla J, Martin-Alvarez PJ, Vara E, Calle JR, Munguira ME, Maranes JP. Comparison between acarbose, metformin, and insulin treatment in type 2 diabetic patients with secondary failure to sulphonylurea treatment. Diabete-Metab 21 ;256-260, 1995.

24. Reaven GM, Lardinois CK, Greenfield MS, Schwartz HC, Vreman HJ. Effect of acarbose on carbohydrate and lipid metabolism in NIDDM patients poorly controlled by sulphonylureas. Diabetes Care 13 (Suppl. 3):32-36, 1990.

25. Rybka J, Gegorova A, Zmydlena A, Jaron P. Clinical study of acarbose. Drug Invest 2: 264-267, 1990.

26. Scheen AJ, Castillo MJ, Lefebvre PJ. Combination of oral antidiabetic drugs and insulin in the treatment of non-insulin-dependent diabetics. Acta Clin Belg 48:259-268, 1993.

27. Coniff RF, Shapiro JA, Seaton TB, Hoogwerf BJ, Hunt JA. A double-blind placebo-controlled trial evaluating the safety and efficacy of acarbose for the treatment of patients with insulin-requiring type II diabetes. Diabetes Care. 18:928-932, 1995.

28. Ledermann H, Hoxter G. Effect of acarbose on postprandial increase in blood glucose. Additive acute effect of once daily administration in insulin treated diabetes. Fortschr Med 122:467-470, 1994.

29. Mertes G, Kiep G. Untersuchung der Wirksamkeit und Verträglicheit von Acarbose über 2 Jahre in Rahmen einer Anwendungsbeobachtung. Basel, 1996 (Abstract).

30. Turner R, Cull C, Holman R, for the United Kingdom Prospective Diabetes Study Group. United Kingdom Prospective Diabetes Study 17: A 9-year update of a randomized, controlled trial on the effect of improved metabolic control on complications in non-insulin-dependent diabetes mellitus. Ann Intern Med 124 (1 pt 2): 136-145, 1996.

31. Holman RR, Cull CA, Turner RC. Glycemic improvement in a double-blind trial with acarbose over one year in 1,946 non-insulin dependent diabetic subjects. Diabetologia 39 (Suppl 1) 156, A44, 1996.

32. Raptis S. European experience in the treatment of IDDM with acarbose in combination with insulin. In: Lefebvre PJ, Standl E eds. New Aspects in Diabetes. Berlin: W De Gruyter 213-226, 1992.

33. Raptis S, Dimitriades G, Karaiskos C et al. Short and long-term studies of acarbose on various metabolic parameters and insulin requirements assessed by the artificial endocrine pancreas. In: Creutzfeldt W, ed. Acarbose, Effects on Carbohydrate and Fat Metabolism. Amsterdam: Excerpta Medica, 393-400, 1982.

34. Walton RJ, Sherif IT, Noy GA, Alberti KGMM. Improved metabolic profiles in insulin-treated diabetic patients given an α-glucosidase inhibitor. Brit Med J 1:220-221, 1979.

35. Dimitriades G, Karaiskos C, Raptis S. Effects of α-glucosidase inhibition on meal glucose tolerance and timing of insulin administration in patients with type 1 diabetes mellitus. Diab Care 14:393-398, 1991.

36. Gerard J, Luyckx AS, Lefebvre PJ. Improvement of metabolic control in insulin-dependent diabetics treated with the α-glucosidase inhibitor acarbose for two months. Diabetologia 1:446-451, 1981.

37. Edmonds ME, Doddryde M, John PN et al. Acarbose and postprandial glycemia in the home-glucose profile. In: Creutzfeldt W ed. Acarbose, Effects on Carbohydrate and Fat Metabolism. Amsterdam: Excerpta Medica 402-405, 1982.

38. Henrichs IA, Heinze E, Kohne E, Teller WM. Improved management of juvenile diabetes by Acarbose. In: Creutzfeldt W ed. Acarbose, Effects on Carbohydrate and Fat Metabolism. Amsterdam: Excerpta Medica 490-493, 1982.

39. Viviani GL, Camogliano L, Borboglio MG. Acarbose treatment in insulin-dependent diabetics. A double blind cross-over study. Curr Therap Res 42:1-11, 1987.

40. McCullogh DK, Kurtz AB, Tattersal RB. A new approach to the treatment of nocturnal hypoglycemia using α-glucosidase inhibition. Diab Care 6:481-487, 1983.

41. Schade DS. Prospective multicenter studies of the efficacy and safety of acarbose (Bay-g-5421) in the treatment of type 1 diabetes mellitus. In: Lefebvre PJ, Standl E, eds. New aspects in Diabetes Mellitus. Berlin: W De Gruyter 237, 1992.

42. De la Calle H, Escobar F, Figuerola D. Clinical efficacy and tolerability of acarbose in IDDM treatment. In: Lefebvre PJ, Standl E, eds. New Aspects in Diabetes Mellitus. Berlin: W De Gruyter 258-261, 1992.

43. Gomez-Perez FJ, Violante R, Sanches-Arriaga LF, Wong B, Rull JA. Efficacy and tolerance to acarbose in non-insulin-dependent diabetics. Rev Invest Clin 44:77-83, 1992.

44. Leonhardt W, Hanefeld M, Fischer S, Schulze J, Spengler M. Beneficial effects on serum lipids in non-insulin-dependent diabetics by acarbose treatment. Arzneim Forsch Drug Res 41:735-738, 1991.

45. Uttenthal LO, Ukponwam OO, Wood SM et al. Long-term effects of intestinal α-glucosidase inhibition on postprandial glucose, pancreatic and gut hormone responses and fasting serum lipids in diabetics on sulphonylureas. Diab Med 3:156-160, 1986.

46. Gerard J, Luyckx AS, Lefebvre PJ. Acarbose in reactive hypoglycaemia: a double-blind study. Int J Clin Pharmacol 22:25-31, 1984.

47. Richard JL, Rodier M, Monnier L, Orsetti A, Mirouze J. Effect of acarbose on glucose and insulin response to sucrose load in reactive hyperglycemia. Diab Metab (Paris) 14:114-118, 1988.

48. Buchanan KD, McLoughlin JC, Alam MJ. Acarbose in the dumping syndrome. In: Creutzfeldt W, ed. Proceedings First International Symposium on Acarbose. Effects on Carbohydrate and Fat Metabolism. International Congress Series 594. Amsterdam: Excerpta Medica, 515, 1982.

49. Qualmann C, Nauck MA, Holst JJ, Orskov C, Creztzfeldt W. Glucagon-like peptide 1 (7-36 amide) secretion in response to luminal sucrose from the upper and lower gut: A study using glucosidase inhibition (Acarbose). Scand J Gastroenterol 30:892, 1995.

50. Ceriello A. Coagulation activation in diabetes mellitus: the role of hyperglycaemia and therapeutic prospects. Diabetologia 36:1119-1125, 1993.

51. Ceriello A, Taboga C, Giacomello R et al. Fibrinogen plasma levels as a marker of thrombin activation in diabetes. Diabetes 43:430-432, 1994.

52. Ceriello A, Taboga C, Tonutti L, Giacomello R, Stel L, Motz E, Pirisi M. Postmeal coagulation activation in diabetes mellitus: the effect of acarbose. Diabetologia 39:469-473, 1996.

53. Spengler M, Cagatay M. The use of acarbose in the primary-care setting: evaluation of efficacy and tolerability of acarbose by postmarketing surveillance study. Clin Invest Med 18:325-331, 1995.

54. Lebovitz HE: Alpha-glucosidase inhibitors. Endocrinology and Metabolism Clinics of North America 26:3, 1997.

55. Rosak C, Nitzsche G, Konig P, Hofmann U. The effect of the timing and the administration of acarbose on postprandial hyperglycemia. Diabetic Medicine 12:979-984, 1995.

56. Gerard J, Lefebvre PJ, Luycckx AS. Glibenclamide pharmacokinetics in acarbose-treated type 2 diabetics. Eur J Clin Pharmacol 27: 233-236, 1984.

57. Scheen AJ, Ferreira Alves de Magalhaes AC, Salvatore T, Lefebvre PJ. Reduction of the acute bioavailability of metformin by the α-glucosidase inhibitor acarbose in normal man. Eur J Clin Invest 24 (Suppl. 3): 50-54, 1994.

58. Hanefeld M, Bär K, Mertes G, Berlinghoff R. Metformin improves metabolic control in non-insulin dependent diabetics with acarbose monotherapy. Diabetologia 40:A312, 1997.

59. Stout RW. Insulin and atherosclerosis. In Stout RW, ed. Diabetes and Atherosclerosis. Dordrecht, the Netherlands: Kluwer Academic Publishers; 165-201, 1992.

Insulin Use in the Elderly

Stephen N. Davis
Jeri B. Brown

SUMMARY. Recent evidence has shown that near-euglycemic control can reduce macrovascular and microvascular complications in individuals with insulin-dependent diabetes mellitus (type 1) and non-insulin-dependent diabetes mellitus (type 2). Large numbers of elderly type 2 patients progress to insulin therapy after oral agent failure. This review outlines treatment goals and age-related considerations for insulin use in the elderly with type 1 and type 2 diabetes. Good glycemic control is a goal of diabetes therapy in the elderly; unfortunately, incidence of severe hypoglycemia increases with tighter control. Reduced counterregulatory responses and awareness to hypoglycemia and advanced atherosclerosis place the elderly at increased risk for morbidity and mortality from hypoglycemia. Consideration must be given to altered drug utilization and use of multiple medications in the geriatric population. Comorbidities, functional impairment, nutritional issues, and age-related learning characteristics present therapeutic challenges. This review describes pharmacodynamics and physiologic effects of available insulin preparations and their application for clinical use in the elderly. Finally, advantages and disadvantages of various regimens used in the

---

Stephen N. Davis, MD, and Jeri B. Brown, RN-C, GNP, are affiliated with the Division of Diabetes and Endocrinology, Vanderbilt University Medical School, Nashville, TN 37221 and the Nashville VA/JDFI Diabetes Research Center, Nashville Veterans Administration Medical Center, Nashville, TN 37212.

This work was supported by grants from the National Institutes of Health (R01DK45369), the Juvenile Diabetes Foundation International, and the Department of Veterans Affairs.

[Haworth co-indexing entry note]: "Insulin Use in the Elderly." Davis, Stephen N., and Jeri B. Brown. Co-published simultaneously in *Journal of Geriatric Drug Therapy* (Pharmaceutical Products Press, an imprint of The Haworth Press, Inc.) Vol. 12, No. 2, 1999, pp. 61-81; and: *Diabetes Mellitus in the Elderly* (ed: James W. Cooper) Pharmaceutical Products Press, an imprint of The Haworth Press, Inc., 1999, pp. 61-81. Single or multiple copies of this article are available for a fee from The Haworth Document Delivery Service [1-800-342-9678, 9:00 a.m. - 5:00 p.m. (EST). E-mail address: getinfo@haworthpress inc.com].

© 1999 by The Haworth Press, Inc. All rights reserved.

elderly with type 1 and type 2 diabetes are discussed. *[Article copies available for a fee from The Haworth Document Delivery Service: 1-800-342-9678. E-mail address: getinfo@haworthpressinc.com]*

KEYWORDS: insulin, aged, hypoglycemia

INTRODUCTION

Prevalence of diabetes in the elderly population (>60 years of age) is extremely high ranging from 15%-33% in Caucasians and African American subjects respectively.[1] Diabetes mellitus (DM) is a complex of syndromes resulting in hyperglycemia and altered nutrient metabolism. Insulin-dependent diabetes mellitus (type 1) occurs as a result of absolute insulin deficiency, and those patients with type 1 diabetes require insulin to sustain life. Non-insulin dependent diabetes mellitus (type 2) is a heterogeneous condition which occurs as a result of deficient β-cell function and insulin resistance. Of all elderly with diabetes, 90%-95% are type 2.[2] Traditional management of type 2 diabetes begins with diet and exercise. Approximately 10% of patients succeed in maintaining desired levels of glycemic control through diet and exercise alone.[3] Progression to oral hypoglycemic therapy follows failure of non-pharmacological measures. In the recent past, only 50% of type 2 patients on oral agents were able to maintain acceptable glycemic control after ten years of treatment.[4] However, with the introduction of new therapies, the duration of adequate glycemic control by oral agents in type 2 patients may be extended. Nevertheless, with the natural history of type 2 diabetes, many patients will require exogenous insulin.[5]

The definitive results of the Diabetes Control and Complications Trial (DCCT) and Kumamato study in Japan demonstrate that improved glycemic control can reduce long-term macrovascular and microvascular complications in individuals with type 1[6] and type 2 diabetes.[7] Thus, the therapeutic goal for all individuals with diabetes should be good glycemic control that often can only be achieved through insulin therapy. Achieving this goal in the elderly poses a problem since many factors that accompany aging complicate management. The purposes of this review are: (1) to review age-related glucose intolerance; (2) to review the physiological action of insulin; (3) to outline special considerations for insulin use in the elderly; and

(4) to consider insulin regimens (including combination therapy with oral agents) for use in elderly type 1 and type 2 patients.

IMPAIRED GLUCOSE TOLERANCE [IGT] WITH AGING

The declining glucose tolerance associated with aging is well-recognized,[8-9] but its direct relationship to aging per se has not been established.[10] Several studies have indicated decreased insulin sensitivity as the age-related mechanism of IGT.[11-14] Impairment of insulin secretion has also been proposed as an age-related mechanism for IGT. Shimizu and co-workers described an alteration in β-cell function in a group of non-obese elderly subjects (n = 17) after 75-g oral glucose load indicating possible predisposition to IGT with normal aging independent of obesity.[15] Others found no significant decrease in insulin secretion in older adults with normal glucose tolerance.[16] Age-related variables such as physical inactivity, use of medications, coexisting illness[17] and delayed carbohydrate absorption[18] may contribute to the deterioration of glucose tolerance with aging.

PHYSIOLOGICAL ACTION OF INSULIN

In non-diabetic man, the endocrine pancreas secretes insulin at a rate of about 40 μg (1 unit [U]) per hour under fasting conditions to act as a tonic inhibitor of endogenous glucose production (EGP), lipolysis and proteolysis.[19] Serum concentrations of 5-15 μU/mL result from this basal secretion. Following stimulation by a meal, peripheral insulin levels increase to concentrations of 60-80 μU/mL returning to basal levels 2-4 hours postprandially.[20]

Insulin acts to regulate glucose homeostasis by: (1) inhibiting glucose production and (2) promoting glucose uptake and storage at liver and muscle and, to a limited extent, adipose tissue. Insulin is released from pancreatic β-cells into the portal vein. In normal man approximately 6% of total insulin release is comprised of proinsulin. In patients with type 2 diabetes, proinsulin can account for 25%-50% of circulating insulin. This is significant as proinsulin has only 5% of insulin's ability to suppress hepatic glucose output and only 3% of insulin's action on skeletal muscle.

Approximately 66% of insulin is cleared by the liver, which establishes a positive portal-peripheral gradient. This physiologic portal-peripheral gradient is reversed during subcutaneous insulin treatment. Consequently, relative hypoinsulinemia can occur in some diabetic patients leading to exaggerated glucose production and hyperglycemia.

Insulin regulates hepatic glucose production by both direct and indirect mechanisms. Insulin can directly inhibit hepatic glycogenolysis and key gluconeogenic enzymes (PEPCK and pyruvate kinase) thereby suppressing hepatic glucose output. However, insulin has potent effects on suppressing lipolysis. An increase of 5 µU/mL of insulin is sufficient to reduce lipolysis by 50%. By reducing lipolysis, insulin limits the flow of non-esterified fatty acids to the liver, thereby inhibiting necessary fuel for gluconeogenesis. In addition, inhibiting lipolysis reduces glycerol output from adipocytes which is an important substrate for gluconeogenesis. These indirect actions of insulin are quantitatively important and may account for 25%-50% of the hormone's role in suppressing glucose production by the liver.

The dose response characteristics of the amount of insulin required to inhibit glucose production by the liver differ in normal and diabetic man. In non-diabetic adults, insulin levels of approximately 20-25 µU/mL will inhibit glucose production by 50%. This degree of insulinemia will also produce quantitatively similar results in lean, metabolically well-controlled type 1 diabetes patients. However, insulin levels of 50-100 µU/mL may be required to reduce endogenous glucose production by 50% in overweight, insulin-resistant type 2 patients.

In non-diabetic adults and adults with type 1 diabetes, insulin levels of approximately 25 µU/mL will also stimulate a 50% increase in glucose disposal. Maximal glucose disposal rates of 10-14 mg/kg/min (basal rate = 2 mg/kg/min) will be achieved at insulin levels of $\cong 100$ µU/mL. Due to insulin resistance, maximal glucose disposal rates in type 2 diabetes patients are only $\cong 4$-6 mg/kg/min and will be obtained at insulinemia > 100 µU/mL.[19]

SPECIAL CONSIDERATIONS FOR INSULIN USE IN THE ELDERLY

Treatment Goals

Glycosylated hemoglobin (HbA1c) of 7.0% (non-diabetic range 4.0%-6.0%) has been recommended as the target for glycemic control in individuals with diabetes. This value translates to average daily plasma glucose levels of ≅140 mg/dL. Recommended fasting and preprandial plasma glucose targets are currently 80-120 mg/dL.[21] There is now definitive evidence that tight glycemic control prevents and/or reduces long-term tissue complications of diabetes. Unfortunately, the price that individuals with diabetes have to pay for good glucose control is an increased incidence of severe hypoglycemia. The consequences of severe hypoglycemia in the elderly can be devastating. Severe hypoglycemia in an individual with compromised cardiac blood flow can result in myocardial infarction and a cerebral vascular event in a person with impaired cerebral circulation. Consequently, in recognition of the increased risk of morbidity and mortality resulting from hypoglycemic episodes in the elderly, blood glucose targets may require relaxation to fasting and preprandial levels ≤ 140 mg/dL and HbA1c of 8.0%. It is also important to consider the many additional factors that may be operating in the elderly (Table 1), and, therefore, therapy needs to be tailored to individual patients.

Good glycemic control is a goal of diabetes therapy in the elderly. Complications of diabetes can be more aggressive in the elderly compared to younger individuals with DM. Therefore, during episodes of acute illness or stress, glycemic control is important to prevent infective exacerbations, fluid and electrolyte imbalance, and renal complications.[22] Short-term use of insulin may be required for: (1) glycemic control during surgery and in the perioperative period, (2) steroid treatment for concomitant illness, and (3) transient periods of severe hyperglycemia. The domicile of an aging individual may fluctuate on a continuum from community dwelling to nursing home residence and, therefore, influence what goals are attainable for an individual patient.

TABLE 1. Special Considerations for Insulin Use in the Elderly

Reduced awareness to symptoms of hypoglycemia
Altered drug utilization and drug interactions
Comorbidities
Functional impairment
Nutritional issues
Age-related learning characteristics

Reduced Awareness of Hypoglycemia

A number of studies have found reduced physiologic responses to hypoglycemia with advanced age. Neuroendocrine (glucagon and growth hormone) and autonomic nervous system (epinephrine) responses are attenuated with advancing age.[23-24] Furthermore, elderly subjects have reduced awareness of autonomic symptoms (e.g., sweating, tremor, palpitations) during hypoglycemia. This is relevant as plasma glucose will fall to lower levels in elderly subjects before corrective action, such as food intake, is instigated. Elderly individuals with diabetes may be at even greater risk for morbidity from hypoglycemia due to superimposed defective hypoglycemic counterregulatory responses and advanced atherosclerosis. Additionally, it should be noted that the incidence of severe hypoglycemia in older persons using insulin is further increased due to inadequate education regarding the signs and symptoms of hypoglycemia.[25]

Altered Drug Utilization and Drug Interactions

Renal blood flow and glomerular filtration rate decline with aging[26] resulting in decreased drug clearance. Impaired renal excretion of insulin increases the potential for hypoglycemia in the elderly. The risk of hypoglycemia limits the use of sulfonylureas in the elderly to agents that are quickly-eliminated, have non-biologically active metabolites and preserve first phase insulin release.[27] It is our practice to use glipizide or glimepiride as first choice-sulfonylurea therapy in the elderly. Hepatic blood flow and parenchymal mass are reduced by approximately 30% by the 8th or 9th decade of life,[28] and hepatic enzyme activity reduces with aging.[29] At least 66% of injected and endogenously-secreted insulin is cleared by hepatic mechanisms. Therefore, subcutaneous insulin and sulfonylurea dosage should be reduced in the elderly. In addition, newer oral agents also cleared by the liver, e.g., troglitazone, will also require careful monitoring for hypoglycemia when used as monotherapy or in combination with insulin. Hypoalbuminemia secondary to acute illness, chronic disease or malnutrition is not uncommon in the elderly. Medications such as glyburide that are highly protein-bound may show a marked increase in unbound drug levels when used alone or in combination with insulin.

Many older adults require multiple medications to manage concomitant chronic illnesses increasing their risk for drug interactions. Between 12 and 17 prescriptions are issued annually to the average elderly adult.[30] Medications frequently used by the elderly population have the potential to increase or decrease the effects of insulin (Table 2). Careful review of medication regimens is essential to prudent management.

Concomitant use of prophylactic anticoagulation and insulin in the geriatric population creates a potentially catastrophic milieu given the inherent risk of hypoglycemia-induced falls. The enigma of falls in the elderly increases their potential for mortality[31] with greater than one-third of individuals over 65 years of age experiencing at least one fall per year.[32] Anticoagulated aging patients on concomitant insulin ther-

TABLE 2. Medications Often Prescribed to the Elderly with Potential for Interaction with Insulin

Drugs which may increase insulin requirements
Glucocorticoids
Lithium
Rifampin
Thiazide diuretics
Progestins
Nicotine
Phenytoin
β-adrenergic receptor agonists
Calcium channel blockers
Clonidine
Morphine
Heparin

Drugs which may decrease insulin requirements
Sulfonylureas
Quinidine
Quinine
Angiotensin converting enzyme inhibitors
Ethanol
Naproxen
Indomethacin
Salicylates
β-adrenergic receptor antagonists

apy require supportive caregivers; thorough education on signs, symptoms and treatment of hypoglycemia; and watchful scrutiny of environmental risks.

Prevalence of musculoskeletal disorders in older Americans is greater than 50%.[33] Palliative therapy for such conditions is often a prescription of non-steroidal antiinflammatory drugs (NSAIDs) with their accompanying risk for renal insufficiency. Risk of renal compromise, especially in the elderly with diabetes, should be minimized with use of alternatives such as acetaminophen or, at minimum, those NSAIDs with shorter half-lives (e.g., ibuprofen, ketoprofen). Corticosteroid treatment of polymyalgia rheumatica and rheumatoid arthritis will alter blood glucose control, and increased insulin doses may be required.

Elderly patients with DM receiving angiotensin-converting enzyme (ACE) inhibitors or β-blockers as antihypertensive therapy have historically been deemed at increased risk for hypoglycemia. Shorr and co-workers recently reported no significant increase in severe hypoglycemia over a 4-year period in 13,599 elderly patients taking insulin or sulfonylureas and antihypertensive agents. They further described no significant difference in incidence of hypoglycemia among patients taking ACE inhibitors, cardioselective or nonselective β-blockers, calcium channel blockers, or thiazide diuretics.[34] Despite the blunting of blood pressure and heart rate response by β-blockers, these agents may paradoxically amplify the hypoglycemic warning sign of diaphoresis.[35] Until further studies clarify the relationship of specific blood pressure-lowering agents to increased risk of severe hypoglycemia, judicious selection of antihypertensives is strongly recommended for elderly patients with DM.

Comorbidities

Elderly patients with type 2 diabetes have increased prevalence of macrovascular disease.[36] A two- to three-fold increased incidence of cardiovascular morbidity is associated with type 2 diabetes.[37] Diabetes mellitus is a risk factor for stroke, peripheral arterial disease and coronary disease,[38] and the elderly have increased susceptibility to severe consequences such as silent ischemia and infarction.[36]

The metabolic syndrome X (insulin resistance, hyperinsulinemia, hypertension, and dyslipidemia) is prevalent in the elderly.[39] Management of insulin resistance of type 2 diabetes requires high doses of insulin. It remains unclear whether the higher doses of insulin required to combat insulin resistance contribute to atherosclerosis and lipid abnormalities.[40] From animal studies it has been suggested that atherogenic effects on the arterial wall may result from hyperinsulinemia.[41] In patients with type 2 diabetes, a positive correlation exists between macrovascular complications and insulin dose;[42] however, causal relationship between insulin dose and macroangiopathy cannot be drawn in such cases since higher doses are required secondary to insulin resistance.[40] Presently there are no compelling data directly linking hyperinsulinemia to increased cardiovascular or cerebrovascular disease in diabetic patients. Epidemiologic studies investigating the relationship of hyperinsulinemia and macrovascular disease fail to partition out the fact that raised insulin levels are merely a surrogate for insulin resistance. Furthermore, as insulin resistance is such a well-recognized component of type 2 diabetes, it is somewhat illogical to attribute the elevated prevalence of atherosclerosis to increased insulin effects.

Many patients with type 2 diabetes are at increased risk for lipid abnormalities (classically ↓HDL and ↑VLDL cholesterol) and obesity.[21] Decreased physical activity and diet consisting of large amounts of saturated fat, unrefined carbohydrates and low fiber content are environmental factors that contribute to the prevalence of type 2 diabetes.[43] Dietary intervention, increased physical activity, and weight loss to combat insulin resistance are important adjunct therapies. Lipid abnormalities in patients with type 2 diabetes should be treated aggressively.[44] The predominant abnormality should determine choice of treatment.

Thyroid disease and diabetes frequently coexist in the geriatric population. Hyperthyroidism; evident as Graves' disease, toxic multinodular goiter, and thyroid adenoma; causes insulin resistance, increases endogenous glucose production and, thus, increases insulin requirements in patients with diabetes. After treatment of thyrotoxicosis, insulin dosages should be readjusted. Hypothyroidism decreases insulin degradation and diminishes appetite mandating adjustment of insulin doses to prevent hypoglycemia.[45] After initiation of thyroid replacement therapy, older patients must be screened for overreplacement, which is deleterious to blood glucose control.

Functional Impairment

Many elderly may be unable to execute complicated insulin regimens secondary to visual impairments, decreased motor coordination, arthritis, or cognitive impairment. Pre-mixed insulins (70/30,50/50), pre-filled syringes, insulin pens[22] and visual magnifying devices are available tools to facilitate insulin administration. Implementation of home health visits to assist elderly by pre-filling syringes is advisable.

Nutritional Issues

Barriers to nutritional management may exist for geriatric patients. Due to limited and/or fixed incomes, many elderly may be unable to purchase food or adhere to prescribed dietary intake. Functional deficits may preclude meal preparation. Adequate assessment and appropriate intervention are essential with insulin use in the elderly.

Age-Related Learning Characteristics

Many older adults demonstrate improved retention of learning when it is self-directed, based on previous experience, and can be applied everyday.[46] These characteristics of older adult learners may facilitate rather than hinder self-management skills necessary for control of DM. Instruction in home blood glucose monitoring (HBGM) enables patients to rapidly adjust pharmacological therapy, physical activity, and dietary intake to prevailing glycemia. HBGM also allows patients to monitor for hypoglycemia and hyperglycemia. Parenthetically, some self blood glucose monitoring devices measure plasma glucose (e.g., Glucometer Elite) while other devices measure whole blood glucose (e.g., One-Touch meters). Plasma glucose levels range from 10%-15% higher than whole blood glucose values. Frequency of monitoring is determined on an individual basis. Understanding of signs, symptoms and appropriate treatment of hypoglycemia including glucagon administration is an integral part of education. In-depth nutritional assessment and counseling are indicated in DM to achieve target levels for blood glucose, promote weight management, and enhance quality of life. Whenever possible, elderly patients should be instructed regarding appropriate dietary choices and portion size to match insulin dose. For those elderly with impairments which prevent learning, caregivers should receive training in all aspects of management.

INSULIN REGIMENS FOR USE IN ELDERLY WITH TYPE 1 AND TYPE 2 DIABETES

The ideal exogenous insulin replacement therapy will simulate the physiological pattern of basal and stimulated insulin secretion. The average daily dose of insulin in type 1 diabetes patients is 0.6 to 0.7 U per kilogram of body weight regardless of age, ethnicity, or length of disease. Requirements of obese patients, both type 1 and type 2, because of insulin resistance are usually greater (2 U/kg/day). Calculation of basal insulin dose is based on fasting blood glucose and body weight and is usually 40% to 60% of the total daily dose. The remainder of the daily dose is given as short-acting insulin in pre-meal doses.[19]

Types of Insulin

Insulin preparations currently available for administration are limited in their ability to mimic the physiological effects of endogenous insulin, but clinical success can be achieved with vigilant management. Insulin preparations can be categorized according to their species of origin and duration of action (Table 3). Porcine and bovine insulins are less frequently used since the advent of human recombinant DNA preparations. Human insulins are preferred for use in the elderly because they provide less antigenicity and improved predictability.[47]

Onset and duration of insulin's action is affected by anatomic injection site, injection technique, smoking, presence of antibodies to insulin, ambient temperature, and differences in individual patient response. Exercise, injection site,[48] and dosage amount cause variations in rate of absorption.

Regular, crystalline insulin is short-acting with onset of action within 30-60 minutes, peaking at 2-3 hours. Its duration is 3-6 hours. Regular insulin may be mixed with NPH, lente, and ultralente insulins. Recommended injection time before a meal is 20-30 minutes.[19]

Lispro insulin is the first human insulin analogue to be approved for

TABLE 3. Insulin Preparations for Use in the Elderly

Type	Onset of action (hours)	Peak of action (hours)	Duration of action (hours)
Short-acting			
Humulin Regular	.5-1.0	2.0-3.0	4.0-6.0
Humulin Lispro	.25	.5-1.5	4.0-5.0
Intermediate-acting			
Humulin N	1.0-4.0	4.0-10.0	14.0-18.0
Humulin L	1.0-4.0	4.0-10.0	16.0-20.0
Humulin Ultralente	6.0-10.0	8.0-20.0	20.0-30.0
Mixed preparations			
Humulin 70/30	0.5-1.0	2.0-12.0	14.0-18.0
Humulin 50/50	0.5-1.0	3.0-10.0	14.0-18.0

Adapted from Davis and Granner.[19]

use by the Federal Drug Administration.[49] Lispro's duration of action is four hours with onset within approximately 15 minutes from injection. Because of its relative stability, lispro can be mixed with human NPH, lente or ultralente,[50] but mixed injections should be given immediately. Absorption rates may change slightly with a mixture of NPH and lispro.[51] Lispro provides more physiologic control of blood glucose compared to regular insulin because of its rapid onset. Risk of hypoglycemia may be reduced with lispro due to its shorter duration of action.[49] Usual timing of lispro injection should coincide with beginning of a meal, but many elderly are erratic eaters or suffer from anorexia secondary to illness or medications. Food intake may be delayed in the aging population given their propensity for more frequent hospitalizations and diagnostic procedures. Lispro can be injected postprandially when timing and amount of intake cannot be easily predicted. Lispro may also be used alone for type 2 diabetes patients to reduce postprandial glucose elevations if endogenous insulin supply remains sufficient to maintain basal coverage.[22]

Intermediate-acting insulins include human NPH and lente. As human-derived insulins, their onset is faster and duration is shorter than the porcine derivatives; thus, pre-dinner injections may not be sufficient to prevent overnight hyperglycemia. The intermediate-acting insulins are usually given once a day before breakfast or twice a day. NPH can be mixed with regular and lispro insulins without any retardation of effect of the more rapid-acting insulins. When mixed, lente may slow the effect of regular or lispro.

Human ultralente has a relatively longer action compared to human NPH or lente and is used to provide basal insulin requirements throughout a 24-hour period. Adjustments in doses, which are given once or twice daily, are made based on fasting blood glucose. Steady state concentrations are not attained for several days after initiation of ultralente.

Insulin is available as pre-mixed combinations which facilitate administration. For the elderly hampered by compromised visual acuity and motor coordination, measuring and drawing up two different insulins is cumbersome. Dosing errors are not uncommon with two-insulin regimens. The two available combinations Humulin 70/30(70% NPH/30% regular) and Humulin 50/50(50% NPH/50% regular) have similar onset and duration of action. The major difference between the two preparations is the amount of regular insulin, and thus, Humulin 50/50 is best used when postprandial hyperglycemia is uncontrolled.

Regimens for Use in Elderly with Type 1 Diabetes

Elderly type 1 patients should be treated with strategies similar to those used in the younger type 1 population with individualized modifications for aging and comorbidities. Multiple dose injections or continuous subcutaneous insulin infusion (CSII) may be used in the elderly. However, it must be remembered that the incidence of severe hypoglycemia is increased three-fold with intensive insulin therapy.[6] If nocturnal hypoglycemia or fasting hyperglycemia occurs in a basal/intermediate, preprandial/short-acting insulin regimen, it is often most efficacious to administer an evening dose of intermediate insulin at bedtime. This maneuver will direct peak insulin action away from the early hours of the morning to pre-breakfast thereby preventing nocturnal hypoglycemia. Furthermore, as peak insulin action is occurring from 4AM to 7AM, fasting hyperglycemia will also be reduced.

Regimens for Use in Elderly with Type 2 Diabetes

Numerous insulin regimens have been tried for treating type 2 diabetes, but controversy continues as to their efficacy.[52] The United

Kingdom Prospective Diabetes Study (UKPDS), a randomized, controlled multi-center trial of 5,102 patients, is currently investigating whether improved glycemic control in type 2 diabetes will prevent complications and whether conventional treatment with diet alone or intensive treatment with metformin, sulfonylurea and insulin is more or less advantageous. After 6 years follow-up, intensive treatment improved HbA1c compared to conventional. After 11 years study duration, results are scheduled for publication in 1998.[53] ADS should provide clearer insight into clinical efficacy of improved glycemic control.

Basal insulin supplement. A minority of type 2 patients may retain sufficient endogenous insulin production so that only basal insulin requirements need to be maintained. For these patients, a single bedtime or two doses of an intermediate-acting insulin given 12 hours apart may be used for glycemic control. However, it should be noted that adequate glycemic control may be elusive even with large doses of insulin.

Combined sulfonylurea and insulin. Combined regimens of sulfonylurea and insulin vary according to timing of insulin dose and choice of sulfonylurea. Patients who are reluctant to begin multiple doses of insulin may accept this approach. In most patients insulin therapy results in weight gain.[40] An increase in central obesity rather than lean body mass has been shown in type 2 patients with initiation of insulin therapy[54] and must be considered when weighing the benefits of improved glycemic control. Obese patients have less weight gain with combination therapy than with insulin alone. The mechanisms responsible for weight gain during insulin or sulfonylurea therapy have been partially elucidated.[55-56] Approximately 66% of weight gain is due to retained calories secondary to reduced glycosuria. Another 10%-15% is due to reduced cellular cycling. The remaining 20%-25% is unexplained but may be due to direct insulin effects on the brain and autonomic nervous system. The amount of weight gained appears to be proportional to prevailing insulin levels. Thus, as it usually requires three-fold greater systemic insulinemia to achieve similarly good glycemic control during exogenous insulin therapy compared to sulfonylurea treatment, it is not surprising that weight gain with insulin therapy is usually far greater than with sulfonylureas.[57-58]

Bedtime insulin, daytime sulfonylurea (BIDS). Unopposed overnight EGP remains a clinical challenge in management of type 2 diabetes. Patients on oral sulfonylurea agents may experience fasting hyperglycemia with adequate glucose control during the day. For these patients a bedtime insulin, daytime sulfonylurea (BIDS) regimen is often implemented.[59] Addition of a bedtime dose of intermediate-acting insulin (NPH or lente) suppresses nocturnal EGP enhancing the daytime action of the sulfonylurea. Substitution of an injection of 70/30 before supper may be indicated if bedtime glycemia is particularly elevated.

Morning insulin combined with daytime sulfonylurea. Daytime sulfonylurea therapy has also been supplemented with morning injection of intermediate-acting insulin. Soneru et al. found this regimen provided glycemic control similar to BIDS therapy; however, incidence of hypoglycemia was increased three-fold with morning insulin.[60]

Split-mixed regimens. A morning dose of mixed regular and intermediate-acting insulin and a second mixed dose before supper has been a frequently-used strategy for nearly three decades. Advantages over a single dose of intermediate-acting insulin include: (1) lower risk of late afternoon hypoglycemia; (2) provision of overnight coverage to improve fasting hyperglycemia; and (3) reduction in postbreakfast hyperglycemia.[47] Drawbacks to this regimen include afternoon hyperglycemia due to lack of fast-acting insulin before midday meal, prolonged hyperinsulinemia due to relatively large doses of intermediate-acting insulin, and nocturnal hypoglycemia.

Multiple dose therapy. Based on HBGM, multiple doses of regular or lispro insulin before meals in addition to intermediate-acting insulin for basal coverage may be required. Initial pre-meal doses should be determined based on usual carbohydrate intake and can be administered as "set" doses. Many community-dwelling elderly prefer to learn carbohydrate counting and adjust pre-meal doses based on eating preferences. This approach allows for greater meal flexibility. Stabilization of basal metabolic control is established with NPH, lente or ultralente after approximately 4 to 5 days. Only adjustments to fast-acting insulin based on HBGM are then needed until basal control is altered.

Insulin pump therapy. Desire for tightened glucose control has brought consideration of insulin pump therapy in type 2 diabetes. Saudek et al. described reduced glycemic variations, hypoglycemia and weight gain in type 2 patients utilizing implantable insulin pump infusion as compared to age-matched patients on multiple daily injections.[61] Although it is unlikely that insulin pump therapy for type 2 diabetes will become routine, it may be useful for highly-motivated and capable patients.

Insulin and biguanide. The biguanide metformin is an oral antihyperglycemic agent that has been used in Europe for over thirty years. Metformin acts by reducing EGP while enhancing peripheral glucose utilization. Studies have shown that metformin may also exert antihyperglycemic effects via non-insulin mediated processes.[62-63] Preliminary data from the FINFAT trial show bedtime insulin and metformin reduced HbA_{1c} values in type 2 patients previously treated with sulfonylurea alone by 2.8% after 1 year with minimal weight gain.[64] Combination insulin and metformin therapy may be useful for obese type 2 patients requiring high insulin doses to alleviate hyperinsulinemia and may improve lipid profiles.[65] Metformin is eliminated mainly through the renal route. Because metformin has no effect on pancreatic insulin secretion and does not cause hypoglycemia, it may be useful in the elderly provided renal, hepatic, and cardiac function are intact.

Insulin and alpha-glucosidase inhibitor. Alpha-glucosidase inhibitors inhibit carbohydrate digestion thereby slowing glucose absorption in the gastrointestinal tract and reducing postprandial glucose levels. Acarbose is currently the only alpha-glucosidase inhibitor available for use in the United States. Flatulence, abdominal cramps, and diarrhea are the most common adverse effects experienced with acarbose use, but symptoms may be reduced with slow dose titration.[66]

In type 2 diabetes patients with uncontrolled postprandial hyperglycemia, a combination of insulin and acarbose offers a reasonable attempt at glycemic control. Okada et al. found hypoglycemic symptoms to be greatly reduced in a group of type 2 patients (3 women and 7 men) after reduction of insulin doses (previously taking 22.6 ± 19.6 U/day) and initiation of alpha-glucosidase inhibitor.[67] Combined insulin and alpha-glucosidase therapy is a viable treatment option to improve quality of life by reducing hypoglycemia in patients previously on large insulin doses.

Insulin and thiazolidinedione. The thiazolidinedione troglitazone became available for use in the United States in 1997. Troglitazone improves insulin sensitivity and, thus, a combination of troglitazone with insulin seems a plausible choice for ameliorating insulin resistance. Improvement in fasting and postprandial hyperglycemia is achieved by troglitazone's enhancement of insulin action on skeletal muscle tissue to increase glucose disposal and at the liver to reduce EGP.[68] In the 991-40 U. S. Registration trial, troglitazone plus insulin provided significant dose-related improvement in fasting plasma glucose and HbA1c in poorly controlled type 2 patients. Liver injury resulting in one liver transplant and five deaths has occurred with troglitazone use. Liver enzymes should be measured at initiation, every month for the first six months, and every other month for the second six months of therapy. Patients should be advised of symptoms of liver dysfunction.

Reduced insulin requirements may manifest after 3-4 weeks of troglitazone therapy. HBGM results should be reviewed frequently by patient and healthcare provider to prevent hypoglycemia.

CONCLUSION

Recent evidence has shown that near-euglycemic control should be advocated in elderly patients with DM to prevent long-term complications. Elderly patients with type 1 diabetes should be managed similarly to younger type 1 patients. Large numbers of elderly type 2 diabetes patients progress to insulin therapy after oral agent failure. Age-related physical changes, comorbidities, multiple medications, socioeconomic issues, and functional impairment present challenges when insulin therapy is needed. Numerous insulin regimens have been proposed without current evidence of which is most efficacious in achieving glucose control without severe hypoglycemia. Selection of insulin regimen, whether as monotherapy or in combination with oral agents, demands individualization with consideration of factors applicable to the elderly population.

REFERENCES

1. National Diabetes Data Group. Diabetes in America (2nd ed.). Bethesda, Maryland: National Institutes of Health; 1995.

2. Morrow, L., and K. Mmaker. Treatment of diabetes mellitus in the elderly. Strategies in Geriatrics, Dept. of Vet. Affairs. 2:1-5;1995.

3. American Diabetes Association. The pharmacological treatment of hyperglycemia in NIDDM. Diabetes Care. 1995;18:1510-1518.

4. Gerich, J. Oral hypoglycemic agents. New Engl. J. Med. 1989;321:1231-1245.

5. Barnett, A. Tablet and insulin therapy in type 2 diabetes in the elderly. J. R. Soc. Med. 1994;87:612-614.

6. Diabetes Control and Complications Trial Group. The effect of intensive treatment of diabetes on the development and progression of long-term complications in insulin-dependent diabetes mellitus. N. Engl. J. Med. 1993;329:977-986.

7. Ohkubo, Y., H. Kishikawa, E. Araki, T. Miyata, S. Isami, S. Motoyosyi, Y. Kojima, N. Furuyoshi, and M. Shichiri. Intensive insulin therapy prevents the progression of diabetic microvascular complications in Japanese patients with non-insulin-dependent diabetes mellitus: a randomized prospective 6-year study. Diabetes Res. Clin. Pract. 1995;28:103-117.

8. Davidson, M. The effect of aging on carbohydrate metabolism: a review of the English literature and a practical approach to the diagnosis of diabetes mellitus in the elderly. Metab. Clin. Exp. 1979;8:688-705.

9. DeFronzo, R. Glucose intolerance and aging. Diabetes Care. 1981;4:493-501.

10. Coordt, M., R. Ruhe, and R. McDonald. Aging and insulin secretion. Proc. Soc. Exp. Biol. Med. 1995;209:213-222.

11. Elahi, D., D. Muller, M. McAloon-Dyke, J. Tobin, and R. Andres. The effect of age on insulin response and glucose utilization during four hyperglycemic plateaus. Experi. Gerontol. 1993;28:393-409.

12. Chen, M., R. Bergman, G. Pacini, and D. Porte, Jr. Pathogenesis of age-related glucose intolerance in man: insulin resistance and decreased β-cell function. J. Clin. Endocrinol. Metab. 1985;60:13-20.

13. Fink, R., O. Kolterman, J. Griffin, and J. Olefsky. Mechanisms of insulin resistance in aging. J. Clin. Invest. 1983;71:1523-1535.

14. Rowe, J., K. Minaker, and J. Pallotta. Characterization of the insulin resistance of aging. J. Clin. Invest. 1983;71:1581-1586.

15. Shimizu, M., S. Kawazu, S. Tomono, T. Ohno, T. Utsugi, N. Kato, C. Ishii, Y. Ito, and K. Murata. Age-related alteration of pancreatic β-cell function. Increased proinsulin and proinsulin-to-insulin molar ratio in elderly, but not in obese, subjects without glucose intolerance. Diabetes Care. 1996;19:8-11.

16. Bourey, R., W. Kohrt, J. Kirwan, M. Staten, D. King, and J. Holloszy. Relationship between glucose tolerance and glucose-stimulated insulin response in 65-year-olds. J Gerontol. 1993;48:M122-M127.

17. Laws, A., and G. Reaven. Effect of physical activity on age-related glucose intolerance. Clin. Ger. Med. 1990;6:849-863.

18. Broughton, D., and R. Taylor. Review: deterioration of glucose tolerance with age: the role of insulin resistance. Age & Ageing. 1991;20:221-225.

19. Davis, S., and D. Granner. Insulin, oral hypoglycemic agents, and the pharmacology of the endocrine pancreas. In: Hardman, J., and L. Limbird, eds. Goodman & Gilman's The Pharmacological Basis of Therapeutics, New York: McGraw-Hill, 1487-1517;1996.

20. Galloway, J., and R. Chance. Improving insulin therapy: achievements and challenges. Horm. Metab. Res. 1994;26:591-598.

21. American Diabetes Association. Standards of medical care for patients with diabetes mellitus. Diabetes Care. 1997;20:S5-S13.

22. Hoogwerf, B., A. Mehta, and S. Reddy. Advances in the treatment of diabetes mellitus in the elderly. Drugs Aging. 1996;9:438-448.
23. Meneilly, G., E. Cheung, and H. Tuokko. Altered responses to hypoglycemia of healthy elderly people. J. Clin. Endocrinol. Metab. 1994;78:1341-1348.
24. Marker, J., P. Cryer, and W. Clutter. Attenuated glucose recovery from hypoglycemia in the elderly. Diabetes. 1992;41:671-678.
25. Mutch, W., and I. Fordyce-Dingwall. Is it a hypo? Knowledge of the symptoms of hypoglycaemia in elderly diabetic patients. Diabet. Med. 1985;2:54-56.
26. Chou, S., and R. Lindeman. Structural and functional changes of the aging kidney. In: Jacobson, H., G. Striker, and S. Klahr, eds. The Principles and Practice of Nephrology. St. Louis: Mosby, 510-514;1995.
27. Jennings, P. Oral antihyperglycaemics: considerations in older patients with non insulin-dependent diabetes mellitus. Drugs Aging. 1997;10:323-331.
28. Lonergan, E., and D. Schmucker. Drug therapy of older patients. In: Katzung, B., ed. Drug Therapy. Norwalk, Connecticut: Appleton & Lange, 15-24; 1991.
29. Piraino, A. Drug use in the elderly: tips for avoiding adverse effects and interactions. Consultant. 1997;37:2825-2834.
30. Moore, A., and M. Beers. Drug interactions in the elderly. Hospital Medicine. 1992;117:684-689.
31. Cohen, H., and J. Crawford. Hematologic problems in the elderly. In: Calkins, E., A. Ford, and P. Katz, eds. Practice of Geriatrics, 541-553;1992.
32. Kustaborder, M. and M. Rigney. Interventions for safety. J. Gerontol. Nurs. 1983;9:159.
33. Sack, K. Musculoskeletal diseases. In: Lonergan, E., ed. Geriatrics. Stamford, Connecticut: Appleton & Lange, 207-235;l 996.
34. Shorr, R., W. Ray, J. Daugherty, and M. Griffin. Antihypertensives and the risk of serious hypoglycemia in older persons using insulin or sulfonylureas. JAMA. 1997;278:40-43.
35. Kerr, D., I. MacDonald, S. Heller, and R. Tattersall. Beta-adrenoreceptor blockade and hypoglycaemia: a randomised, double-blind placebo-controlled comparison of metoprolol CR, atenolol, and propranolol LA in normal subjects. Br. J. Clin. Pharmacol. 1990;29:685-693.
36. American Diabetes Association. Implications of the Diabetes Control and Complications Trial. Diabetes Care. 1997;20:S62-S64.
37. Garcia, M., P. McNamara, T. Gordon, and W. Kannell. Morbidity and mortality in diabetics in the Framingham population: sixteen year follow-up study. Diabetes. 1974;23:105-111.
38. Vokonas, P., and W. Kannel. Diabetes mellitus and coronary heart disease in the elderly. Clinics in Geriatric Medicine. 1996;12:69-78.
39. Muller, D., D. Elahi, J. Tobin, and R. Andres. The effect of age on insulin resistance and secretion: a review. Semin. Nephrol. 1996;16:289-298.
40. Koivisto, V. Insulin therapy in type II diabetes. Diabetes Care. 1993;16:29-39.
41. Taskinen, M. Hyperlipidemia in diabetes. Baillieres Clin. Endocrinol. Metab. 1990;4:743-775.

42. Janka, H., A. Ziegler, E. Standl, and H. Mehnert. Daily insulin dose as a predictor of macrovascular disease in insulin-treated non-insulin-dependent diabetics. Diabetes Metab. 1987;13:359-364.

43. Bennett, P. Epidemiology of non-insulin-dependent diabetes. In: LeRoith, D., S. Taylor, and J. Olefsky, eds. Diabetes mellitus: a-fundamental and clinical text. Philadelphia: Lippincott-Raven, 455-459; 1996.

44. Havel, R., and E. Rapaport. Management of primary hyperlipidemia. N. Engl. J. Med. 1995;332:1491-1496.

45. Shetty, K., and E. Duthie. Thyroid disease and associated illness in the elderly. Clinics in Geriatric Medicine. 1995;11:311-322.

46. Gessner, B. Adult education: the cornerstone of patient teaching. Nursing Clinics of North America. 1989;24:589-595.

47. Beaser, R. Fine-tuning insulin therapy. Postgraduate Medicine. 1992;91: 323-330.

48. American Diabetes Association. Insulin administration. Diabetes Care. 1997; 20: S46-S49.

49. Bohannon, N. Benefits of lispro insulin. Postgraduate Medicine. 1997;101: 73-79.

50. Campbell, R., L. Campbell, and J. White. Insulin lispro: review of its role in the treatment of diabetes mellitus. Ann. Pharmacother. 1996;30:1263-1271.

51. White, J., R. Campbell, and I. Hirsch. Insulin analogues: new agents for improving glycemic control. Postgraduate Medicine. 1997;101:58-66.

52. Yki-Jarvinen, H., M. Kauppila, E. Kujansuu, J. Lahti, T. Marjanen, L. Niskanen, S. Rajala, L. Ryysy, S. Salo, P. Seppala, T. Tulokas, J. Viikari, J. Karjalainen, and M. Taskinen. Comparison of insulin regimens in patients with non-insulin-dependent diabetes mellitus. N. Engl. J. Med. 1992;327:1426-1433.

53. U. K. Prospective Diabetes Study Group. U. K. Prospective Diabetes Study 16: overview of 6 years' therapy of type II diabetes: a progressive disease. Diabetes. 1995;44:1249-1258.

54. Sinha, A., C. Formica, C. Tsalamandris, S. Panagiotopoulos, E. Hendrich, M. DeLuise, E. Seeman, and G. Jerums. Effects of insulin on body composition in patients with insulin-dependent and non-insulin-dependent diabetes. Diabetic Medicine. 1995;13:40-46.

55. Flier, J. Obesity and lipoprotein disorders. In: Kahn, R., and G. Weir, eds. Joslin's Diabetes Mellitus. Philadelphia: Lea & Febiger, 351-356;1994.

56. Bray, G. Basic mechanisms and very low calorie diets. In: Blackburn, G., and G. Bray, eds. Management of Obesity by Severe Caloric Restriction. Littleton, Mass: PSG, 129-169;1995.

57. Nathan, D., A. Roussell, and J. Godine. Glyburide or insulin for metabolic control in non insulin dependent diabetes mellitus. Ann. Int. Med. 1988; 108:334-340.

58. Henry, R., B. Gumbiner, T. Ditzler, P. Wallace, R. Lyon, and H. Glauber. Intensive conventional insulin therapy for type II diabetes. Diabetes Care. 1993; 16:21-31.

59. Genuth, S. Insulin use in NIDDM. Diabetes Care. 1990;13:1240-1264.

60. Soneru, I., Agrawal, J., Murphy, A., et al. Comparison of morning or bedtime insulin with and without glyburide in secondary sulfonylurea failure. Diabetes Care. 1993; 16:896-901.

61. Saudek, C., Duckworth, W., et al. Implantable insulin pump vs multiple-dose insulin for non-insulin-dependent diabetes mellitus: a randomized clinical trial. JAMA. 1996;276:1322-1327.

62. Klip, A., and L. Leiter. Cellular mechanism of action of metformin. Diabetes Care. 1990; 13:696-704.

63. Wu, M., P. Johnston, W. Sheu, C. Hollenbeck, C. Jeng, I. Goldfine, Y. Chen, and G. Reaven. Effect of metformin on carbohydrate and lipoprotein metabolism in type 2 diabetes patients. Diabetes Care. 1990;13:1-8.

64. Yki-Jarvinen, H., K. Nikkila, I. Ryysy, T. Tulokas, R. Vanamo, and M. Heikkila. New thoughts on insulin therapy in Type 2 diabetes. 16th International Diabetes Federation Congress. Abstracts of the State-of-the Art Lectures and Symposia; 1997.

65. Klepser, T., and M. Kelly. Metformin hydrochloride: an antihyperglycemic agent. Am. J. Health Syst. Pharm. 1997;54:893-903.

66. Tan, G., and R. Nelson. Pharmacologic treatment options for non-insulin-dependent diabetes mellitus. Mayo Clin. Proc. 1996;71:763-768.

67. Okada, S., K. Ishii, H. Hamada, S. Tanokuchi, K. Ichiki, and Z. Ota. Can alpha-glucosidase inhibitors reduce the insulin dosage administered to patients with non-insulin-dependent diabetes mellitus? Journal of International Medical Research. 1995;23:487-491.

68. Suter, S., J. Nolan, P. Wallace, B. Gumbiner, and J. Olefsky. Metabolic effects of new oral hypoglycemic agent CS-045 in NIDDM subjects. Diabetes Care. 1992;15:193-203.

Index

Acarbose, 29-31,41-42,76-77
   coagulation and, 54
   drug interactions of, 55
   European experience with, 47-56
   GLP-1 secretion and, 53
   with insulin in type 1 diabetes, 51-52
   with insulin in type 2 diabetes, 50-51
   lipid effects in type 2 diabetes, 52-53
   in place of diet therapy, 48-49
   in place of other oral agents, 49-50
   in reactive hypoglycemia, 53
   tolerability of, 54
ACE inhibitors, 68
Aging
   body composition in, 8-11
   drug interactions and, 65-67
   drug utilization and, 65
   functional impairment in, 70
   glucose homeostasis in, 11-16
   insulin therapy in, 64-71
   learning characteristics in, 70-71
   nutrition in, 70,72
   pathophysiology in, 5-17
Alpha-glucosidase inhibitors, 47-57. *See also* Acarbose
American Diabetes Association (ADA)
   glycemic control guidelines, 26-27
Anticoagulants, 67-68
Antihypertensives, 68
AWP (average wholesale price) of drugs, 39-40

Beta-blockers, 68
Biguanides, 14,76. *See also specific drugs*
Body composition, 8-11

Coagulopathy, 54
Combination therapy, 37,41-42,50-52, 74-77
Comorbidities, 68-69
Complications
   glycemic control and, 23-27,65
Cost of therapy, 39-41

Diabetes Control and Complication Trial (DCCT), 23-24,48,62
Diabetes mellitus type 1
   acarbose in, 51-52
Diabetes mellitus type 2
   genetic factors in, 7
   incidence and prevalence of, 7-8
   insulin therapy in, 63-77. *See also* Insulin therapy; Insulin
   management in elderly, 21-43. *See also* Management
   pathophysiology of, 5-17
Dietary therapy
   acarbose with, 48-49
Drug interactions, 55,65-67
Drug utilization, 65
Dumping syndrome, 53

Functional impairment, 70

GLP-1 (glycoliprotein-1), 53
Glucose intolerance
   aging and, 7-8
Glucose levels
   diagnostic, 6
   fasting and aging, 8
Glucose monitoring, 70-71
Glucose toxicity, 16-17
Glucose transporter system, 13
α-Glucosidase inhibitors, 47-57. *See also* Acarbose

Health care system, 40-41

Hepatic glucose production, 13-14
Hyperinsulinemia, 13-14,69
Hyperlipidemia, 52-53,69
Hypoglycemia
   reactive, 53
Hypoglycemic episodes, 24,37-38,65, 66,73
Hypothyroidism, 69

Impaired glucose tolerance (IGT), 13-15,63
Insulin
   physiological action of, 63
   types of, 71-73
   visceral fat and, 9-12
Insulin pump, 75
Insulin resistance, 68-69
   aging and, 11-12
   syndrome of (syndrome X), 8, 22-23,68-69
Insulin secretion
   failure in, 14-15
Insulin therapy, 36-39,61-78
   acarbose with, 50-52
   cost of, 39-41
   discontinuance of, 38-39
   potential problems with, 37-38
   principles of, 36-37,62-63
   special considerations in aging, 64-71
   in type 1 diabetes, 73
   in type 2 diabetes, 73-77
Islet-cell antibodies, 7

Kumamato study, 62

Learning characteristics, 70-71
Lipid levels. *See* Hyperlipidemia
Lispro insulin, 71-72

Management, 21-43
   choice of treatment regimen, 41-42
   control and complications, 23-26
   cost issues in, 39-41
   guidelines for glycemic control, 26-27

insulin therapy, 36-39,61-78. *See also* Insulin therapy
oral therapy, 27-36. *See also* Oral therapy *and specific agents*
Metformin, 14,31-33,41-42,50,76
Mortality, 25

Nephropathy, 24,25,68. *See also* Complications
Neuropathy, 25-26
NIDDM (type 2 diabetes mellitus). *See* Diabetes mellitus type 2
NSAIDs, 68
Nutritional management, 70,72

Obesity, 8-11,15,49. *See also* Visceral fat
Oral therapy, 27-36
   acarbose, 29-31,41-42,47-56
   drug interactions, 55
   metformin, 14,31-33,41-42,59,76
   sulfonylureas, 28-29,41-42,50, 74-76
   troglitazone, 34-36,41-42

Pathophysiology, 5-17
Patient education, 70-71
Prevention, 16

Reactive hypoglycemia, 53
Retinopathy, 24,25. *See also* Complications

Self-care, 3
Sulfonylureas, 28-29,41-42,50,74-76
Syndrome of insulin resistance (syndrome X), 8,22-23, 68-69

Thrombosis, 54
Thyroid disease, 69
Triglycerides. *See* Hyperlipidemia
Troglitazone, 34-36,41-42

Visceral fat, 9-10

Whole-body glucose utilization, 11-12
Wisconsin Epidemiologic Study, 25

AN INFORMATIVE NEW BOOK!

Come to grips with the current problems with the medication use process and discover ways to improve it!

IMPROVING THE QUALITY OF THE MEDICATION USE PROCESS

Over 200 Pages!

Error Prevention and Reducing Adverse Drug Events

Edited by Alan Escovitz, PhD
Director, Pharmacy Extension Services, College of Pharmacy and Director, Center for Continuing Health Sciences Education, The Ohio State University; Exective Director, Council of Ohio Colleges of Pharmacy

Dev S. Pathak, DBA
Director of the Center for Health Outcomes, Policy, and Evaluation Studies (HOPES); Merrell Dow Professor in the Division of Pharmacy Practice and Administration at The Ohio State University

Philip J. Schneider, MS, RPh
Director, Ambulatory Pharmacy Education and Service and Associate Professor, Pharmacy Practice and Administration, The Ohio State University

"AN EXCELLENT INTRODUCTION TO A SERIOUS PUBLIC HEALTH PROBLEM.... This dialogue among health professionals is an important step in understanding the complex human and systems failures that lead to error."
—Donald A. Goldmann, MD, Medical Director of Quality Improvement, Children's Hospital, Boston, Massachusetts

Explores the nature and significance of problems with the medication use process, regulatory control over the process, and ways to improve the process.

Selected Contents
Medication Misadventuring as a Public Policy Issue • Danger at the Drug Store • Adverse Drug Events • A Regulatory Response to the Occurrence of Adverse Drug Events • The State Boards of Pharmacy Response to the Occurrence of Adverse Drug Events • Practitioner Experiences as a Catalyst for Change • Assessing the Performance of the Medication Use Process • Health Professions Responsibility for Reducing Adverse Drug Events • Pharmacist's Role in Reducing Adverse Drug Events • Nurse's Role in Reducing Adverse Drug Events • Reducing Adverse Drug Events • Role of Information Systems in Reducing Adverse Drug Events • Re-Engineering the Medication Use Process in an Organized Health Care System • Reengineering a Pharmacy Practice to Reduce Adverse Drug Events • Index • Reference Notes Included

$49.95 hard. ISBN: 0-7890-0458-5.
(Outside US/Canada/Mexico: $60.00)
Text price (5+ copies): $29.95.
(Outside US/Canada/Mexico: $36.00)
1998. Available now. 228 pp. with Index.
Includes tables/figures, a bibliography, and Web site/Internet addresses.

VISA, MASTERCARD, DISCOVER, AMERICAN EXPRESS & DINERS CLUB WELCOME!

CALL OUR TOLL-FREE NUMBER: 1–800–HAWORTH
US & Canada only / 8am—5pm ET; Monday–Friday
Outside US/Canada: +607–722–5857

FAX YOUR ORDER TO US: 1–800–895–0582
Outside US/Canada: +607–771–0012

FACULTY: For an examination copy of **Improving the Quality of the Medication Use Process** on a 60-day examination basis, see order form on back page. You will receive an invoice payable within 60 days. **If you decide to adopt the book, your invoice will be cancelled.** (Offer available only to teaching faculty in US, Canada, and Mexico.)

E-MAIL YOUR ORDER TO US:
getinfo@haworthpressinc.com

VISIT OUR WEB SITE AT:
http://www.haworthpressinc.com

Pharmaceutical Products Press®
An Imprint of The Haworth Press, Inc.
10 Alice Street, Binghamton, NY 13904–1580 USA

BIC99

TO ORDER: CALL: 1-800-HAWORTH / FAX: 1-800-895-0582 (outside US/Canada: + 607-771-0012) / E-MAIL: getinfo@haworthpressinc.com

☐ YES, please send me Improving the Quality of the Medication Use Process
___ in hard at $49.95 ISBN: 0-7890-0458-5. (Outside US/Canada/Mexico: $60.00)

- Individual orders outside US, Canada, and Mexico must be prepaid by check or credit card.
- Discounts are not available on 5+ text prices and not available in conjunction with any other discount.
- Discount not applicable on books priced under $15.00.
- 5+ text prices are not available for jobbers and wholesalers.
- Postage & handling: in US: $4.00 for first book, $1.50 for each additional book.
  Outside US: $5.00 for first book; $2.00 for each additional book.
- NY, MN, and OH residents: please add appropriate sales tax after postage & handling.
- Canadian residents: please add 7% GST after postage & handling.
- If paying in Canadian dollars, use current exchange rate to convert to US dollars.
- Please allow 3-4 weeks for delivery after publication.
- Prices and discounts subject to change without notice.

Signature _____

☐ BILL ME LATER($5 service charge will be added).
(Minimum order: $15.00. Bill-me option is not available for individuals outside US/Canada/Mexico. Service charge is waived for jobbers/wholesalers/booksellers.)
☐ Check here if billing address is different from shipping address and attach purchase order and billing address information.

☐ PAYMENT ENCLOSED $ _____
(Payment must be in US or Canadian dollars by check or money order drawn on a US or Canadian bank.)

☐ PLEASE BILL MY CREDIT CARD:
☐ Visa ☐ MasterCard ☐ American Express ☐ Discover ☐ Diners Club

Account Number _____

Expiration Date _____

Signature _____

May we open a confidential credit card account for you for possible future purchases?
( ) Yes ( ) No

THE HAWORTH PRESS, INC., 10 Alice Street, Binghamton, NY 13904-1580 USA

Please complete the information below or tape your business card in this area.

NAME _____

INSTITUTION _____

ADDRESS _____

CITY _____

STATE _____ ZIP _____

COUNTRY _____

COUNTY (NY residents only) _____

E-MAIL _____
May we use your e-mail address for confirmations and other types of information?
( ) Yes ( ) No. We appreciate receiving your e-mail address and fax number. Haworth would like to e-mail or fax special discount offers to you, as a preferred customer. We will never share, rent, or exchange your e-mail address or fax number. We regard such actions as an invasion of your privacy.

☐ YES, please send me Improving the Quality of the Medication Use Process (ISBN: 0-7890-0458-5) to consider on a 60-day examination basis. I understand that I will receive an invoice payable within 60 days, or that if I decide to adopt the book, my invoice will be cancelled. I understand that I will be billed at the lowest price. (Offer available only to teaching faculty in US, Canada, and Mexico.)

Signature _____

Course Title(s) _____

Current Text(s) _____

Enrollment _____

Semester _____ Decision Date _____

Office Tel _____ Hours _____

(13) 01/99 BIC99